GATTEGNO MATH

TEXT-BOOK 5

The Text-Book Series

1. Qualitative arithmetic.
   The study of numbers from 1 to 20.

2. Study of numbers up to 1000.
   The four operations.

3. Applied arithmetic.

4. Fractions, Decimals, Percentages.

5. Study of numbers.

6. Applied mathematics.

7. Algebra and geometry.

GATTEGNO

MATHEMATICS

TEXT-BOOK 5

by

C. GATTEGNO

EDUCATIONAL SOLUTIONS
NEW YORK, N.Y. 10003

© *1970 C. Gattegno*
SBN 87825 015 8

The title GATTEGNO MATHEMATICS
embodies an approach best expressed by
the phrase The Subordination of Teaching
to Learning. The programme covered in
this series envisages the use of colored
rods (ALGEBRICKS) and other books
and printed materials that are obtainable
from:

EDUCATIONAL SOLUTIONS INC.
821 Broadway, New York, N.Y. 10003

Printed in Great Britain by
Lamport Gilbert Printers Ltd.,
Wantage Road, Reading, England

CONTENTS

I

THE FORMATION OF NUMBERS

PART I

THE FORMATION OF NUMBERS

Place value and powers of ten

1. If you take a handful of rods and put them end-to-end, you can find the length thus obtained. It is the sum of the lengths of the composing rods.

In fact, you do not need to put the rods end-to-end. You can, *by inspection*, find the answer and there are many ways of doing it. Take a few handfuls and see which way you think is the best according to the composition of each handful.

2. Going back to the idea of putting rods end-to-end, you can see it is a way of obtaining lengths which can be increased on and on by simply adding a rod to the length you already have. Could you find a length which is bigger than any other?

If you think so, write it down.

Once you have written it down, think of that length and add one more rod to it. Is this new length bigger than the one you had? Can you therefore say that there is a largest number?

3. Only a few of the numbers have names you can remember, but numbers can always be given a name and there are methods of telling the name of any number written down and of writing any number whose name is given.

We learnt in Book 2 to read and write any number of a reasonable length. We shall now reconsider the matter using the towers of orange rods with maybe one other rod on top.

Let us start with a tower made of four orange rods. Its name is ten thousand or 10,000; if we now replace the top rod by any other rod—red, light green, pink, . . . blue—we obtain

2

what we read as two, three, four, . . . nine thousand and write in figures as 2,000; 3,000; 4,000; . . . 9,000.

Now, it is clear that if we make a 'five tower' of, say, four orange rods and one black, so that it represents 70,000, we can equally well regard it as representing 7×10 for the top two rods and 1,000 for the bottom three. So 70,000= $70 \times 1,000$ and also $1,000 \times 70$. We can extract other similar calculations from this tower such as:

$7 \times 10,000$	and	$10,000 \times 7$
700×100	and	100×700
$7,000 \times 10$	and	$10 \times 7,000$

or we can find more factors than two:

$$7 \times 100 \times 100$$
$$7 \times 10 \times 1,000$$
$$70 \times 10 \times 100$$

and so on.

Make a tower of your own with four or five rods and write down all the ways you can find of reading it as multiplications, as we have done above.

4. Imagine now that you make several towers; each has orange rods with another rod on top. Since you can read them separately and can write them down, you can now learn to write the answer to their addition.

For example, you know what 7,000, 300, 20 look like as crosses or towers. $7,000 = 70 \times 100$, so if we add 7,000 and 300, we get 73 hundred, which we shall read as seven thousand and three hundred, and write as 7,300.

But $7,300 = 730 \times 10$ or 730 tens; if we add to it 20, or 2×10, we get 732 tens which we shall write as 7,320 and read as seven thousand three hundred and twenty.

Now, if we recall that a pink could be put end-to-end with that length, giving us 24 at the end instead of 20, we find that 7,324 is the way to write that length as measured by the white one; and 'seven thousand three hundred and twenty-four' is the way to read it.

3

Of course, there are many ways of reading a number, but only a few have become current; others can be used for fun, but they are not considered useful in general. 7,324 can be read seventy-three hundred and twenty-four. This is particularly used for dates above 1000, when we say, for example, 'eighteen hundred and fifteen' for 1815, shortening it sometimes to 'eighteen fifteen'.

Read in one or more ways the following numbers and write down the words for them:

5,617	1,066	3,001	7,044	9,999	4,683
4,444	7,800	6,403	7,315	2,942	1,696

Largest and smallest number of n figures

5. Which is the largest number of four figures? which is the smallest?

Write them both down and find the difference between them.

Write numbers of five figures and read them; write the words you say and compare them with those your neighbor has found.

Which is the largest number of five figures? and which is the smallest?

Compare the largest number of four figures with the smallest of five; what is the difference between these two numbers?

Write down the names of the following numbers:

32,715	13,654	41,611	40,200	71,900
55,555	23,456	49,888	56,680	77,700
64,000	93,903	60,000	71,001	49,999

And of the following:

143,774	506,007	617,291	320,320	567,890
300,009	400,400	999,999	701,700	610,000

Write down the smallest number of seven figures and compare it with the largest of six figures. What is the difference between these two numbers?

4

6. A million is the name for the 'six tower' made of orange rods. As it can be made of two "threes" on top of each other, we can write $1,000 \times 1,000 =$ one million or 1,000,000.

We can play with this tower, which represents a million, in the way we did in Section 3, writing down the various multiplications to be found in it.

Complete the pattern in writing, using your rods if you cannot do it:

$1,000,000 = 100 \times$	$1,000,000 = 10 \times$
$1,000,000 = 100 \times 100 \times$	$1,000,000 = 10 \times 10 \times 10 \times$
$1,000,000 = 10,000 \times$	$1,000,000 = 100,000 \times$

Any number of seven figures is bigger than 1,000,000 (unless it is equal to it) and smaller then ten million or 10,000,000. By using the same method as above, read the following seven figure numbers and write down what you are saying:

2,739,192	4,000,385	4,350,000	5,001,000
1,111,111	6,543,210	9,019,000	8,305,609
3,120,140	9,999,999	8,900,000	7,777,123

Index notation

7. By building taller towers using orange rods, we can go on meeting larger numbers. The 'seven tower' made of orange rods is called ten million and is written as 10,000,000; the 'eight' called one hundred million is written as 100,000,000; the 'nine' is called one thousand million (in the U.S.A. a billion) and is written as 1,000,000,000. In Great Britain, a billion means a million million and therefore requires twelve zeros to express it in writing.

Using the *index notation* for the towers, or powers of ten, we can form the following table:

No. of rods in Tower	Number expressed in figures	Written as a power
1 $10 = 10$		10^1 (ten)

5

2	$10 \times 10 = 100$	10^2
		(one hundred)
3	$10 \times 10 \times 10 = 1,000$	10^3
		(one thousand)
4	$10 \times 10 \times 10 \times 10 = 10,000$	10^4
		(ten thousand)
5	$10 \times 10 \times 10 \times 10 \times 10 = 100,000$	10^5
		(one hundred thousand)
6	$10 \times 10 \times 10 \times 10 \times 10 \times 10 = 1,000,000$	10^6
		(one million)
7	$10 \times 10 \times 10 \times 10 \times 10 \times 10 \times 10 = 10,000,000$	10^7
		(ten million)
8	$10 \times 10 \ldots$ (8 times) $= 100,000,000$	10^8
		(one hundred million)
9	$10 \times 10 \ldots$ (9 times) $= 1,000,000,000$	10^9
		(one thousand million)
10	$10 \times 10 \ldots$ (10 times) $= 10,000,000,000$	10^{10}
		(ten thousand million)

After one million we see how we form the name of large numbers. If we write, for example, 10,000,000,000,000 with 13 zeros, we read it as ten million million, which can also be written 10^{13}.

Read 10^{22}, 10^{17}, 10^{25}, 10^{11}.

8. These towers made of orange rods can be used as in No. 4 to show how to read any number. In fact, we count the figures in a number and compare this in our table with the name that goes with that number, and for each figure we have a name.

Thus 73,449,786 is equal to:

73,000,000, or seventy-three million, plus
    449,000, or four hundred and forty-nine thousand, plus
        786, or seven hundred and eighty-six—

in all, seventy-three million four hundred and forty-nine thousand, seven hundred and eighty-six.

6

This reading must become second nature, so make sure you can read any number that can be put down in figures.

Horizontal and vertical notations

9. In operations with large numbers, most people need to write them down, though there are some people who enjoy remembering huge numbers and operating on them in their heads. Find out how big a number you can remember after naming it or looking at it for a few moments. Find out also how easily you can add large numbers in your head.

For those who need the help of paper and pencil, the rules for writing down operations before performing them safely are simple and were already met in Book 2.

When you write numbers you wish to add or subtract, put the figures that have the same name directly underneath each other. By the 'same name' we mean *units*, *tens*, *hundreds*, etc.:

Thus:

7	17	5	29	102	491	47	2092
+8	+ 5	+17	+32	+345	+ 47	+491	+ 118

and

4632	5631	932170	171	4	
− 57	− 1473	− 160406	+6093	+7935	etc.

Write vertically the following operations:

(1) 7,043,216+43,241+2,417+443,611+389+98+ 61,732,973

(2) 4,219+346+91,972+4,600+473,602+3,195,701+8

(3) 73,441,663+69,173,298

(4) 17,002,008−8,324,219

and find the answers by whatever method you choose.

Equivalent subtractions

10. Let us consider what can be done with subtractions of large numbers (which we know how to simplify as we discovered in Book 2). For example, in the following subtraction,

$$73,485$$
$$-68,697$$

we can consider the various difficulties we have to overcome. If we worked out the subtraction from the right, since all figures of the second number, except the last, are bigger than the corresponding one above, we cannot subtract the figures of the same name. Methods have been found to help perform the subtraction in spite of that obstacle. Here we shall use the principle we met in the subtraction of smaller numbers, i.e. that each subtraction belongs to a family of equivalent ones which contain some examples that are very easily worked out.

If we add three to each of the numbers given above, we make an equivalent subtraction, like this:

$$73,488$$
$$-68,700$$ Add 300 to each to get: $$73,788$$
$$-69,000$$ and the answer

to this is obviously 4,788.

Another way would have been to subtract from the left: 68 from 73 leaves 5. So we can replace the given subtraction by the equivalent one:

$$5,485$$
$$-\ 697$$ If we add 303 to each we get $$5,788$$
$$-1,000$$ which gives

4,788 as the answer.

There are several other ways of finding the answer. Try your hand at the following, giving as many solutions as you can by adding or subtracting convenient numbers in each case.

2,001	13,202	45,625	10,008
−1,986	− 9,877	−39,846	− 9,989

Growth of numbers

11. The following questions will enable you to think about numbers in a way that will give you a deeper knowledge of them. Go slowly and understand fully what each question means, because in that way you will gain much more.

When you add ten numbers of seven figures each, which is the smallest number you can obtain? Which is the largest?

If you add numbers of six figures, how many must you add to make sure that the answer has eight figures? Could you obtain such a result if you added half that number of numbers? What could you then say of the numbers added?

How can one get larger and larger numbers?

If you had to obtain larger and larger numbers by addition, would you choose to go on adding ones or tens or hundreds?

12. The following questions are about multiplication and the growth of numbers by multiplication.

If you multiply a number of one figure by another of one figure, how many figures are there in the answer?

Which is the smallest number we can get by multiplying two numbers of one figure? Which is the largest?

If we multiply a number of two figures by a number of one, how many figures are there in the answer? Which is the smallest and which is the largest we can get by these multiplications?

If we multiply two numbers of two figures, how many figures are in the answer? Give the smallest and the largest number you can obtain in that way.

Multiplication speeded up

13. In multiplication, we call the number that is multiplied the *multiplicand*, the other number the *multiplier ;* the

result is the *product*. Often we use the word product for any number of numbers separated by the sign ×, though the resultant number has not been worked out. Example: 7 × 8 is the product of 7 and 8.

If we exchange the rôle of multiplicand and multiplier, do we change the result? How can you make sure that your answer is true in all cases?

In multiplications by the methods we explained in Book 2, we use a step-by-step operation ending up by an addition. For example, in 326 × 42, we first multiply 326 by 2, then 326 by 40 and add the results. In writing, we use the pattern

$$\begin{array}{r} 326 \\ \times 42 \\ \hline 652 \\ 13{,}040 \\ \hline 13{,}692 \end{array}$$

Calculate the number of operations you actually must undertake in order to find the answer in the following cases:

The multiplicand has one figure, the multiplier two.

The multiplicand has two figures, the multiplier two, three or four.

14. Let us compare in space, speed and satisfaction the following operations:

   42 × 8   and   42+42+42+42+42+42+42+42

which give the same result.

Write an addition of seven equal numbers of 5 figures. Time yourself and compare it with multiplying that number by 7.

Which would you prefer, to add 136 equal numbers of 3 figures or to multiply that number by 136?

How many operations must be carried out to find the product of two numbers of three figures each?

Find the number of operations in the multiplication of two numbers having, respectively,

2 and 3 figures each, 3 and 4, 2 and 4, 2 and 6, 3 and 5, 4 and 5, 2 and 10.

Do you save time if you multiply the smaller by the larger or the converse? Can you count the saving in the number of operations?

Do you think there is any other consideration to take into account besides the number of figures?

Which of the following multiplications would you prefer to perform if you agree that the multiplier is written second?

$$\begin{array}{cc} 79,864 & 222,222 \\ \times 222,222 & \text{or} & \times 79,864 \end{array}$$

Find out by experience which is the easier and shorter, and say why.

Practice will help you to decide which of two multiplications is the simpler.

15. In this section we shall introduce a suggestion you may like to adopt. You need not do so, because we have included it only for those who find it fun to work with numbers.

If we know the products of numbers with two figures, in addition to the products of numbers with one figure, we can greatly extend our power and can become very quick at long multiplications and divisions.

To gain experience of these products, and to remember them, is easier than it seems if we go the right way about it. We need to gain a feeling for the pattern of numbers so that if we remember one part of the pattern it instantly suggests other connected parts. When we discovered, for instance, that $5 \times 8 = 40$ and that $8 \times 5 = 40$, these ideas became connected in our minds so that both were more readily remembered. We can use this principle when we meet larger numbers.

11

One way of doing this is by *duplication*, by which is meant that we take a number such as 7 and double repeatedly. In this way we obtain useful milestones:

7   14   28   56   112   224   448   896   1792

These numbers are all products of 7 and another number. 14 is the product of 7 and 2, and 28 is the product of 7 and 4 and of 14 and 2. If we play with these numbers, breaking them down into factors, we get:

$$7 \times 16 = 112$$
$$14 \times 16 = 224 = 7 \times 32$$
$$28 \times 16 = 448 = 14 \times 32 = 7 \times 64$$
$$56 \times 16 = 896 = 28 \times 32 = 14 \times 64$$
$$112 \times 16 = 1792 = 56 \times 32 = 28 \times 64$$

We soon find that a pattern takes shape in our minds, so that these numbers and their relation to one another become fixed in the memory.

By doing the same thing with 6, 8 and 9, we obtain many more products, and we can then begin to fill in the gaps left in our knowledge.

As soon as we have a sufficient stock of products and factors we can work out in one line the answer to such multiplications as 1754×2345. We shall need to know the following products:

$$17 \times 45 = 765, \qquad 17 \times 23 = 391$$
$$23 \times 54 = 1242, \qquad 45 \times 54 = 2430$$

We can see what we have to do if we first analyse the multiplication we have to do in this way:

$$1754 \times 2345$$
$$= (1700 + 54) \times (2300 + 45)$$
$$= 17 \times 23 \times 10,000 + (17 \times 45 + 54 \times 23) \times 100 + 54 \times 45$$

We shall see the structure of the operation we are doing if we study this diagram:

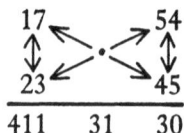

12

Here is the working:

$45 \times 54 = 2430$. Put down 30 on the right and carry 24
$17 \times 45 = 765$ and $54 \times 23 = 1242$. Add 765, 1242 and the 24 carried, which makes 2031. Put down 31 and carry 20.

$17 \times 23 = 391$. Add 20 carried and put down 411 on the left of 3130.

The answer is, thus, 4,113,130, and as we have done the working in our heads it will look like this on paper:

$$\begin{array}{r} 1{,}754 \\ \times 2{,}345 \\ \hline 4{,}113{,}130 \end{array}$$

This can only be done if we know the products required and can also do the carrying and adding in our heads.

Now let us say that we have to multiply in one line, as quickly as possible, the following large numbers:

$$351{,}754 \times 192{,}345$$

We already know that these numbers can be written as follows:

$$351{,}754 = 35 \times 10{,}000 + 17 \times 100 + 54$$
$$192{,}345 = 19 \times 10{,}000 + 23 \times 100 + 45$$

To find their product we must know by heart:

54×45, 17×45 and 23×54, 35×45 and 19×54, 35×23 and 19×17.

If we know them we can write down the answer in one line, doing the multiplying, the carrying and the additions in our heads.

The answer, before actually working it out, looks like this:

$35 \times 19 \times 100{,}000{,}000 + (35 \times 23 + 19 \times 17) \times 1{,}000{,}000 + (35 \times 45 + 54 \times 19 + 17 \times 23) \times 10{,}000 + (17 \times 45 + 54 \times 23) \times 100 + 54 \times 45$.

A diagram will, once again, help us to see the structure of the operation and, this time, dots are added which will help.

We work from right to left, using them to lead us from each separate operation to the next:

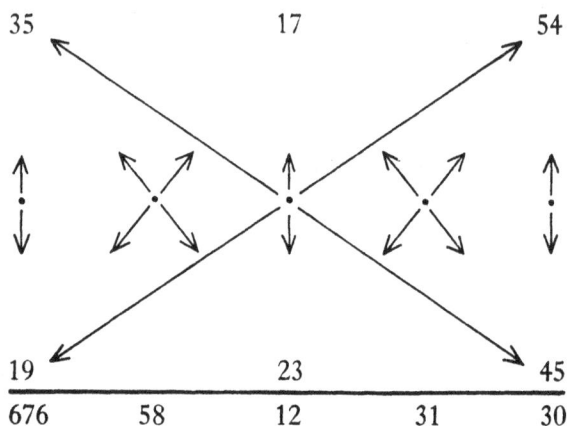

35		17		54
19		23		45
676	58	12	31	30

This, written in the ordinary way, is:

$$67,658,123,130$$

Working:

$54 \times 45 = 2430$ Put down 30 and carry 24

$17 \times 45 = \ \ 765$
$54 \times 23 = 1242$
Carried $\quad \ \ 24$
___
$\qquad \quad 2031$ Put down 31 and carry 20

$35 \times 45 = 1575$
$54 \times 19 = 1026$
$17 \times 23 = \ \ 391$
Carried $\quad \ \ 20$
___
$\qquad \quad 3012$ Put down 12 and carry 30

$35 \times 23 = \ \ 805$
$17 \times 19 = \ \ 323$
Carried $\quad \ \ 30$
___
$\qquad \quad 1158$ Put down 58 and carry 11

14

$$35 \times 19 = 665$$
Carried   11
―――
676   Put down 676

Time yourself when you do these multiplications, first in the new and then in the old way, and see how you progress with practice.

You can, of course, put down any large numbers you want to multiply together and find which products you need to know by heart before starting. Only when you know them is it fair to time yourself, for otherwise you will waste much time finding the intermediate products.

If you practise this you will be able to do these large multiplications so rapidly that people will think you are a prodigy.

Growth of numbers

16. You have seen that when these two numbers of six figures are multiplied together, the answer is a number of eleven figures. Most people will consider these numbers as large, though we know that we can go on and on making larger numbers.

Multiplication is an operation that takes us in big leaps further and further along the number series. But there are much faster procedures we can use.

We shall now look at a few of these that are not too difficult.

Start with 2 and go on doubling; in how many steps do you obtain a number which is bigger than,

100 or   1,000
10,000 or   100,000
1,000,000 or   10,000,000?

Now take the successive powers of 3, 4, 5, 6, 7, 8, 9, by making towers with the light green, pink, . . . blue rods, finding the answer each time. How many storeys would you need to build to find an answer bigger than a million?

15

This can be written like this:

Find n so that 3^n is bigger than 1,000,000

Find p ,, ,, $4^p > 10^6 = 1,000,000$

Find q ,, ,, $5^q > 10^6$,,

Find r ,, ,, $6^r > 10^6$,,

Find t ,, ,, $7^t > 10^6$,,

Find u ,, ,, $8^u > 10^6$,,

Find v ,, ,, $9^v > 10^6$,,

17. Can you reach or exceed any number if you add a given number a certain number of times to itself?

Suppose A is the large number you have chosen, can you exceed that number if you start with a red rod and add red rods to it one after another? Is it quicker if you start with a black rod, adding black rods?

In writing, it would look like this:

If A is the large number chosen and a is a given number; can you always find n so that

$$n \times a > A?$$

Suppose you have written a number of 15 or 20 figures; can you find a multiple of 2 or 3, or 5 or 9 or 11 bigger than that number?

How many operations must you perform in order to get a number larger than 173,249 if,

(a) you start with 2 and *add* 1 every time?

(b) you start with 2 and *add* 10 every time?

(c) you start with 2 and *multiply* by 2 every time?

(d) you start with 10 and *multiply* by 2 every time?

(e) you start with 10 and *multiply* by 3 every time?

(f) you start with 10 and *multiply* by 10 every time?

18. If we now consider towers made from a different coloured rod each time, we can get the values of

2^n, 3^n, 4^n, 5^n, 6^n, 7^n, 8^n, 9^n, 10^n

for values of n equal to 1, 2, 3, 4, etc.

Compare the results with each other when n has the same value for each tower.

Can you always say that n is $<2^n$? (The sign $<$ is the opposite of $>$ and means *less than*).

When you have understood how to find the answer to a^n if $a = 2$, or 3, or 4 . . . or 10 and $n = 1$ or 2 or 3 etc.,

find $(1\cdot5)^n$ if $n=1$, 2, 3, 4, 5, etc. stopping when you wish.
$(1\cdot2)^n$ if $n=1$, 2, 3, 4, 5, etc.   ,,   ,,   ,,   ,,
$(1\cdot1)^n$ if $n=1$, 2, 3, 4, 5, etc.   ,,   ,,   ,,   ,,

Now compare the results.

After how many multiplications is

$(1\cdot5)^n > n$? i.e. is $(1\cdot5)^2>2$, $(1\cdot5)^3>3$ etc.?
$(1\cdot2)^n > n$? i.e. is $(1\cdot2)^2$ or $(1\cdot2)^3$ or $(1\cdot2)^4$ bigger than the *index*?
$(1\cdot1)^n > n$? i.e., as before.

A *fixed* number taken to different powers is an example of what is called *exponentiation*, a new word to add to addition, multiplication and duplication with which we are already familiar.

19.   There is yet another way of getting large numbers quickly, but it will require closer attention than the previous ones.

If we make staircases in writing, using a succession of the *same* number, we can see that the successive steps are

$$2 \quad 2^2 \quad 2^{2^2} \quad 2^{2^{2^2}} \quad 2^{2^{2^{2^2}}} \quad \text{etc.}$$

The first is 2, the next 4, the next $2^4=16$, the next $2^{16}=256$, and the next $2^{65,536}$ which is too big for us to calculate.

Using 10 instead of 2, we have the staircases:

$$10 \quad 10^{10} \quad 10^{10^{10}} \quad 10^{10^{10^{10}}} \quad \text{etc.}$$

and already the third, $10^{10^{10}}$, has ten thousand million figures in ordinary notation.

17

Compare what we find if we begin with the number 10 and take three steps, using each of the methods we have considered:

$$10+3=13$$
$$10\times3=30$$
$$10^3 \quad =1,000$$

$$10^{10^{10^{10}}}=\text{a number too large to calculate}$$

20. To sum up what we have found in this study of obtaining larger and larger numbers, we can say that we understand what

$$a+N, \quad a\times N \quad a^N, \quad a^{a^{a^{a^{a^{a^a}}}}}$$

mean and that, if a and N are bigger than 1, and N is a whole number, all these numbers are bigger than a.

If $N=4$ and $a=2$, find the four answers
If $N=2$ and $a=4$, find the four answers.

What is the effect of exchanging N and a in the four operations?

The experience you have gained in this part of the book may give you an enthusiasm for numbers and greatly extend your power to deal with them.

II

DIFFERENT BASES OF NUMERATION

PART II

DIFFERENT BASES OF NUMERATION

Polynomials

1. In all your work so far you have had the set of rods containing the ten different colors and have formed your numbers, measuring the lengths with orange rods. In Book 2 you used the towers of rods to form large numbers. In your mind a tower of seven orange rods across each other with a red one on top is read 20,000,000 or 2×10^7. A tower with six orange rods and one black one on top, will form 7,000,000 or 7×10^6. Placing these towers near each other and thinking of them as to be added, we have a representation of the number 27,000,000. Similarly, to represent the number 27,638,496 we need some further towers until we finally reach a blue by orange cross and a dark green rod. This shows that we could write for the number 27,638,496 the expression:

$$2 \times 10^7 + 7 \times 10^6 + 6 \times 10^5 + 3 \times 10^4 + 8 \times 10^3 + 4 \times 10^2 + 9 \times 10 + 6.$$

This we shall call a *polynomial in* 10; the highest power being 7, we shall say its *degree* is 7; 10 being the notational base we have used, we shall call it the *argument*; write the following numbers as polynomials in 10, giving their degree:

43,597
209,703
6,679
7,689,364
741
37,400
29

2. Now, any number for us is also a length, and we measure that length using the orange rods. We know that

we do a similar thing to find the cardinal number of a set by counting, since what we do is to use the unit repeatedly; when we measure the corresponding length with the white rods, we again do a similar thing.

In this section of this book we are going to learn to measure any length using any rod or combination of rods, and then to write or read any such arrangement of rods.

First, can we make any length using only:

white rods?

red rods? or do we need red and white?

green rods? or do we need green and red? or green and white? or green and red and white?

pink rods? which others *must* we have to make with the pink rods any length?

Answer the same questions for each colored rod in turn.

Binary scale

3. Let us take the red rods and the white ones only. Can we make any length with them?

We shall make the convention that we shall never use two whites when one red will do. Let us write the names of the lengths of all the rods, using for their measurement the red and white rods, and denoting by $+$ the operation of end-to-end.

At the same time we can write these names in a notation involving only 1's and 0's, in which a 1 represents the presence of a tower, cross or rod, and a 0 represents the absence of such a tower, cross or rod, as shown in the following table.

white	$=w$	one white	1
red	$=r$	one red, no white	10
l. green	$=r+w$	one red, one white	11
pink	$=r+r$	one cross red-red, no red, no white	100
yellow	$=r^2+w$	one cross red-red, no red, one white	101
d. green	$=r^2+r$	one cross red-red, one red, no white	110

21

black	$=r^2+r+w$	one cross red-red, one red, one white	111
tan	$=r^3$	one "three" red, no cross, no white	1000
blue	$=r^3+w$	one "three" red, no cross, one white	1001
orange	$=r^3+r$	one "three" red, one red, no white	1010

It is clear that we can go on using only red and white rods, and go on writing in notation only 1's and 0's for any length. Work out a few examples of your own, and check whether the following are correct:

| two orange rods end-to-end | r^4+r^2 | or | 10,100 |
| three tan rods end-to-end | r^4+r^3 | or | 11,000 |

Conversely, what is the length that is represented by 11,101 or 10,101?

4. It is clear that we can form a polynomial using r as the argument. If we write such a polynomial in the following way, choosing the degree to be 7, we have:

$$Ar^7+Br^6+Cr^5+Dr^4+Er^3+Fr^2+Gr+H$$

then, A, B, C, ... H can only equal 1 or 0; but A cannot be a 0 otherwise the degree is less than 7. For each choice of the values of these capital letters we obtain one length only.

Let us try a few examples:

(a)   A=1, B=1, C=0, D=0, E=1, F=0, G=1, H=1.

Using the argument r the number will be written 11,001,011, and the length will be made of one white, one red, no pink, one tan, eight tans, sixteen tans end-to-end (or in cm. and using the decimal scale it would be 203).

In our red towers each time we introduce a new rod we double the number and we know that r^3 equals the tan, r^4 equals two tans and r^6 the cross tan-tan, r^7 will be 16 tans or the "three" red-tan-tan.

(b)   A=0, B=1, C=1, D=1, E=0, F=0, G=0, H=1.

22

This number will be written 1,110,001, and the length for which it stands will be made of: one white, no red, no pink, no tan, two tans, four tans, eight tans end-to-end, (or in cm. it would be written as 113, using the decimal scale). As $A=0$ in this example we have a polynomial of degree 6.

(c) $A=1$, $B=0$, $C=1$, $D=0$, $E=1$, $F=0$, $G=1$, $H=0$.

This number will be written 10,101,010, and will represent the length in cm. (base decimal) equal to 170.

5.   Let us have a table of the powers of 2 so that we can easily change any number given in the *decimal notation* into the notation using only 1's and 0's, called the *scale of 2* or the *binary notation*.

$r^0\ r^1\ r^2\ r^3\ r^4\ \ r^5\ r^6\ \ r^7\ \ r^8\ \ r^9\ \ r^{10}\ \ldots$ with $r=2$ gives
1 2 4 8 16 32 64 128 256 512 1024

Now, given any number between, say, 1 and 2,000, we can change it into a sum of powers of 2. Make sure of that and work out the following examples as we do the scheme for 1,873 as follows:

$1,873=1024+849$, $849=512+337$, $337=256+81$,
$81=64+17$ $17=16+1$
So $1,873=r^{10}+r^9+r^8+r^6+r^4+r^0$

This is a polynomial in r of degree 10, which translated into our scale of 2 is: 11,101,010,001. Find the binary forms for:

$$739$$
$$1,497$$
$$1,006$$
$$1,702$$

Scale of three

6.   If instead of the red rods we use the light green rods, we know that we can, if we form polynomials (or mental sum of towers containing green, red and white rods), obtain any number and write it in the scale of 3. Here too we must only

use one red or one white rod in order to complete the lengths measured by the light green rods.

Let us again measure the rods white to orange, this time using green, red and white rods only, and the notations 0, 1, 2 for them

white	=no light green, no red, one white	1
red	=no light green, one red, no white	2
light green	=one light green, no red, no white	10
pink	=one light green, no red, one white	11
yellow	=one light green, one red, no white	12
dark green	=a cross red-green, no red, no white	20
black	=a cross red-green, no red, one white	21
tan	=a cross red-green, one red, no white	22
blue	=a cross green-green, no red, no white	100
orange	=a cross green-green, no red, one white	101
orange + white	=a cross green-green, one red, no white	102

Form a few more lengths, and write them in the new notation.

Check whether the following are correct:

Two orange rods end-to-end	—	202
Three orange and a dark green	—	1,100
Five blue and a black	—	1,221

7. The polynomials we can form with g as the argument are written as follows:

$$Ag^{10}+Bg^9+Cg^8+Dg^7+Eg^6+Fg^5+Gg^4+Hg^3+Ig^2+Jg+K$$

We have taken it of degree 10, and the capital letters take the values 0, 1 or 2 only.

Let us write some of these for special choices of the capitals and find the lengths they represent in decimal notation and measured in cm.

24

(a) A=0, B=2, C=0, D=1, E=0, F=2, G=1,
H=0, I=2, J=2, K=1.

In the scale of 3 the number is written 2,010,210,221, and its value in the decimal system is 42,145, as can be found by putting the corresponding lengths end-to-end.

(b) A=1, B=0, C=1, D=0, E=2, F=0, G=2,
H=1, I=0, J=1, K=0.

Then umber is written, in scale of 3 notation, 10,102,021,010, and the length it represents in cm. is 67,260 using the decimal notation.

All this can be easily found by having at one's disposal a table of the powers of 3:

g^0	g^1	g^2	g^3	g^4	g^5	g^6
1	3	9	27	81	243	729

g^7	g^8	g^9	g^{10}	
2,187	6,561	19,683	59,049	. . .

Let us see what we did in order to find the value in the decimal scale, of the numbers written in the scale of 3. From the table of powers of 3 we took the corresponding numbers (or the double of them if so indicated), and added them up. Is not this that we did in examples (a) and (b) above?

8.   The table above can be used for the transformation of numbers given in decimal notation into ones in the scale of 3.

For example:

$$7389 = 6561 + 828, \quad 828 = 729 + 99, \quad 99 = 81 + 2 \times 9$$

so, $7389 = g^8 + g^6 + g^4 + 2g^2$ or, in scale of 3: 101,010,200.

Find similarly the corresponding writing in that scale of the decimal

40,029
15,678
8,040
4,384
1,712
1,000
100
70
54

Scale of seven

9. What we know how to do in the scale of 2, 3 and 10 could be done in any scale. To get more practice we shall now work out the same schemes with one scale smaller than 10, say 7, and with one larger, say 12.

To work in the scale of 7 we shall need the rods up to the black, and shall use the following table:

b^0	b^1	b^2	b^3	b^4	b^5	b^6	b^7
1	7	49	343	2,401	16,807	117,649	823,543

We shall form polynomials in b with the value of the capital letters being any of the numbers 0, 1, 2, 3, 4, 5, 6.

$$Ab^7 + Bb^6 + Cb^5 + Db^4 + Eb^3 + Fb^2 + Gb + H$$

In the following examples we have in the scale of 7:

(a) A=0, B=0, C=1, D=4, E=6, F=1, G=5, H=2 or 146,152, equal, in the scale of 10, to 28,555. Is it true?

(b) A=1, B=2, C=0, D=0, E=6, F=4, G=3, H=6 or 12,006,436, equal, in the scale of 10, to 1,061,122. Is it true?

As before, find the expressions in the scale of 7, of the following numbers written in the scale of 10:

34,600	1,000	72	10
256,197	200	20	9
51,401	120	19	8

26

Scale of twelve

10. In order to be able to write in the scale of 12, we shall need two more signs in addition to the ten figures 0, 1, 2 ... 9 used in the scale of 10. Let us use X and Y as the additional signs. We can form the table of the powers of 12 to work with. We form polynomials in 12, with the capital letters (A, B, C etc.) being allowed to assume the value of X or Y, as they were allowed to assume the values 0, 1, 2 ... 9.

Perhaps the following dictionary of the first few numbers in the two scales will help us to become familiar with the new system which has more than the usual 10 figures:

Scale 10	Scale 12	Scale 10	Scale 12	Scale 10	Scale 12
1	1	10	X	19	17
2	2	11	Y	20	18
3	3	12	10	21	19
4	4	13	11	22	1X
5	5	14	12	23	1Y
6	6	15	13	24	20
7	7	16	14	25	21
8	8	17	15	26	22
9	9	18	16	27	23

and so on.

Form your table of the powers of 12, and write polynomials such as:

$$A(12)^7 + B(12)^6 + C(12)^5 + D(12)^4 + E(12)^3 + F(12)^2 + G(12) + H$$

You may form polynomials of degree 7 as the one above, or of any other degree; then give to A, B, ... H, values as in the following examples:

A=1, B=0, C=X, D=3, E=0, F=Y, G=2, H=1

or A=2, B=4, C=7, D=X, E=Y, F=X, G=3, H=X

or A=0, B=0, C=Y, D=9, E=7, F=0, G=X, H=0

Is it true that for these three examples we have the writings 10 X 30 Y 21, or 247 XYX 3 X, or Y 970 X 0, respectively? Having made sure, find their corresponding numbers in the scale of 10.

Find the expressions in the scale of 12, for the following decimal forms:

32	4,481	94,007
411	7,200	201,409
1,700	18,289	1,456,783

Change of base

11. To be able to change the form of a number given in any scale into its form in the scale of 10 and conversely, is now an easy, if not a very short, operation. Can we transcribe numbers directly when given in the scale of 7 to the form in the scale of 8, for example, without passing through the scale of 10?

Let us start with small numbers.

56 scale 7, is equal to $5 \times 7 + 6$ or 5 black rods + 1 dark green.

But this length is made alternatively of 5 tan and a white, so that:

56 (scale 7)=51 (scale 8).

So, if we have the rods we can write a given length in one scale and measure it using the *base* rod of the other scale in which we wish to transcribe the length.

Using your rods, obtain the following transformations of small numbers:

(scale 6)   32   45   21   53   10   13
    into scale 7, 8 and 9.

(scale 9)   81   37   40   72   64   28
    into scale 4, 5, 7.

You may not wish to use your rods for this. In that case you may have to work out mentally in the scale of 10. Write down what you are doing and express the mechanism of your operations as a rule, so that you can save time when applying it. Then apply it to the transformation of any two numbers in thesc ale of 2 into their forms in the scale of 7, and conversely.

The four operations in any scale

12. We are going to consider now the four operations in any base, and to see that the rules we use for the scale of 10, are valid for all the other scales.

If we add two numbers written in the scale of 2, we have to carry a unit every time the result adds up to 2 or more. For example:

$1,010+101,010=110,100$	$1,010$	$11,101$
$11,101+101,101=1,001,010$	$+101,010$	$+101,101$
	$\overline{110,100}$	$\overline{1,001,010}$

In the scale of 3 we carry a unit when the result is equal to or over 3. For example:

$12,021$	Find the	$1,020,102$	$2,102,210$
$+211,210$	answers to	$+12,122,120$	$+11,222,012$
$\overline{1,001,001}$			

In the scale of 7 we carry a unit when the result is equal to or over 7. For example:

$543,216$	Find the	$24,632$	$123,456$
$+423,645$	answers to	$+54,554$	$+654,321$
$\overline{1,300,164}$			

Similarly, in a subtraction, say in the scale of 6:

$$234,521$$
$$-142,454$$
$$\overline{52,023}$$

In the scale of 8 find the answer to:

$$407,561$$
$$-364,672$$

For multiplication too we use exactly the same method as in the scale of 10. If we form a multiplication table, say in the scale of 7, we find:

$1\times1=1$	$1\times2=2$	$1\times3=3$	$1\times4=4$	$1\times5=5$	$1\times6=6$
$2\times1=2$	$2\times2=4$	$2\times3=6$	$2\times4=11$	$2\times5=13$	$2\times6=15$

29

$3\times1=3$	$3\times2=6$	$3\times3=12$	$3\times4=15$	$3\times5=21$	$3\times6=24$
$4\times1=4$	$4\times2=11$	$4\times3=15$	$4\times4=22$	$4\times5=26$	$4\times6=33$
$5\times1=5$	$5\times2=13$	$5\times3=21$	$5\times4=26$	$5\times5=34$	$5\times6=42$
$6\times1=6$	$6\times2=15$	$6\times3=24$	$6\times4=33$	$6\times5=42$	$6\times6=51$

This table can be used to multiply any two numbers in the scale of 7. The following examples are to be read using the tables above and carrying up figures as usual.

```
    24            124            412
  × 23          × 423          × 324
  -----         -----          -----
   105            405           2251
    51            251           1124
  -----           532           1536
   615          -----          -----
                56,415        200,421
```

Check these multiplications by finding in the scale of 10, the numbers that you multiply, as well as the answers.

13. If we write two numbers in the scale of 10 as polynomials in 10, for example:

$$(a\times10^3+b\times10^2+c\times10+d) \times (e\times10^4+f\times10^3+g\times10^2+h\times10+i)$$

after multiplication and collection of the same powers of 10 we get:

$ae\times10^7 + (af+be)\times10^6 + (ag+bf+ce)\times10^5 + (ah+bg+cf+de)\times10^4 + (ai+bh+cg+df)\times10^3 + (ch+dg+ib)\times10^2 + (ci+dh)\times10+di.$

This can be found also with polynomials of any argument, 7 or 12, for example. Since we know that two numbers can be written as polynomials of any argument, their product is also a polynomial of the same argument. Try it out for numbers written in the scale of 6, for example.

When the numbers in the bracket of the product-polynomial are greater than 10, then there is a carry to a higher power. This also happens when we work in a base different from 10, as you can see in the examples at the end of No. 12.

14. Division can also be performed in any base in the same way as it is done in the base of 10. For example, remaining in the scale of 7, let us divide 252 by 5. Dividing 5 into

250 we know that $5\times30=210$, so 40 will be the remainder; bringing 2 down we have to divide 5 into 42, that gives 6.

Try out any divisions you like in different
scales and observe the similarity of the
operations. Since we know by heart the
tables in the scale of 10 we find it easier to
perform divisions in that base than in
others. That will no longer be the case if we have at our disposal tables like the one on the previous page.

```
      36
5)252 | 30
   42 | 6
```

15. The advantages of working with bases other than 10 are best felt when using *computers* or machines that can do all kinds of calculations. For example, the *electronic machines* use the scale of 2, and for that system the table of multiplications reduces to the following products:

$0\times0=0$, $1\times0=0\times1=0$, and $1\times1=1$

What do we do to multiply two numbers in this system? For instance, 1101×101 is written:

```
      1101
    × 101
     ─────
      1101
   110100
   ───────
  1000001
```

So, in fact it reduces merely to an addition.

Check this multiplication in the scale of 10.

Because we only have to add units and shift one place to the left every time we add two units, the only difficulty is in the translation of the numbers written in the base 10, into their form in the scale of 2. If that were done in advance, then all operations would become additions.

For example $7^2=7\times7$, this product in the scale of two is 111×111 or

31

$$\begin{array}{r} 111 \\ \times 111 \\ \hline 111 \\ 1110 \\ 11100 \\ \hline 110{,}001 \end{array}$$

110,001 is written $2^5 + 2^4 + 1$ and equals 49 in the scale of 10.

Later on in your studies of mathematics or science, you will meet descriptions of calculating machines including electronic ones making systematic use of the scale of 2.

III

DIVISIBILITY
AND PRIME NUMBERS

PART III

DIVISIBILITY AND PRIME NUMBERS

Eratosthenes sieve

1. We have already met *prime numbers* in various contexts. We can find all those that are smaller than a given number if we use a device called *Eratosthenes sieve* because its inventor, who lived in Alexandria some 200 years B.C., compared it with the device used to separate sand from stones.

Write all the numbers up to, say, 1,000 and cancel all multiples of 2 except 2, all multiples of 3 except 3, all multiples of 5 except 5, etc. Those that are left are the prime numbers smaller than 1,000. Of course, it is quicker to write 2 down and no other even numbers, then 3 and no multiple of 3, 5 and no multiple of 5, etc.

2. Count the prime numbers in every decade, in every group of fifty numbers, in every century, in every five centuries. Make a table:

NUMBER OF PRIMES

per decade		per fifty		per century	
1—10	....	1—50	....	1—100	....
11—20	....	51—100	....	101—200	....
21—30	....	101—150	....	201—300	....
31—40	....	151—200	....	301—400	....
.. ..	....	.. ..	....	.. ..	....
.. ..	....	.. ..	....	.. ..	....

Do you notice any regularity?

3. Are there many examples of pairs of consecutive prime numbers between 1 and 1,000? Example 11, 13; 17,

34

19 . . . Give them all. Do you find that they are distributed in any regular way?

4. Prime numbers are those that have no factors except 1 and themselves, and in that they resemble each other. But we can distinguish them in many ways.

Thus $11=2\times5+1$.

Can you find another prime that is a multiple of 5 to which 1 is added?

Also $13=3\times4+1$.

Can you find another prime which is a multiple of 4 to which 1 is added?

Can you find primes that are of the form: $1 +$ multiple of 8.

5. The following results are obviously true:

$$5=4+1 \quad 13=9+4 \quad 17=16+1 \quad 29=25+4 \quad 37=36+1$$

On the right hand sides the two numbers are squares of numbers which are not necessarily prime themselves. Find a few primes which have that same property of being equal to the sum of two squares.

Have they got something in common?

6. $2^{2^1}+1=5 \quad 2^{2^2}+1=17 \quad 2^{2^3}+1=257$

Is 257 in your table of primes?

Can you calculate $2^{2^4}+1$? Is it a prime number?

Criteria of divisibility

7. We know that all even numbers are divisible by 2, and when we study the written form of a number we can easily tell whether it is divisible by 3 or 5. It is not so easy to see whether it is divisible by 11, but there is a way which can be readily understood and which gives us a simple test. Here are the steps that lead us to the test:

We find that 10, 100, 1,000 and 10,000 etc: are either one more or one less than a number that *is* divisible by 11:

$10=11-1$; $100=99+1$; $1,000=1,001-1$; $10,000=9,999+1$, and so on. The subtraction and addition of 1 goes alternately as we proceed from 10 to 100 to 1,000 and upwards.

Let us choose a number, say 32,594, and make use of what we have found to test whether it is divisible by 11. We can write this number in several different ways, which lead on from one another:

(i) $3\times10,000+2\times1000+5\times100+9\times10+4$

(ii) $3\times(9999+1)+2\times(1001-1)+5\times(99+1)+9\times(11-1)$ $+4$

(iii) (a multiple of 11)$+3+$(a multiple of 11)$-2+$(a multiple of 11)$+5+$(a multiple of 11)$-9+4$

(iv) (a multiple of 11)$+(3+5+4)-(2+9)$

(v) (a multiple of 11)$+1$.

It is clear that (a multiple of 11)$+1$ is not divisible by 11 and, accordingly, 32,594 is not divisible by 11.

So the test we have obtained to check whether a number is divisible by 11 is: *the difference between the sum of the figures in the even places and the sum of the figures in the odd places must be* 0 *or a multiple of* 11 *if the number itself is divisible by* 11.

Write down any number at random. Check whether it is divisible by 11. If it is not, what alteration to it must you make to produce a number that is?

The *criteria for divisibility* in the case of the other prime numbers either cannot be found or are too difficult to be of any use to us.

Finding prime factors

8. In order to know whether a given number is prime, we are left with the method of Eratosthenes or of trying in turn to divide the number by all primes that are smaller than this number. If it is divisible by a smaller number, we obtain factors of the given number.

If the following numbers are not prime, find their factors:

117	143	247
257	1,761	3,721

Note that if a number, for example, equals 17×19, when 17 is tried out, the other factor is obtained at the same time; but a number may be a product of several primes. After the first prime is found, we have to repeat the same procedure with the other factor.

For instance, $18 = 2 \times 9 = 2 \times 3 \times 3$.

This last form of 18 is called the form reduced to its *prime factors*.

Here is an example of working out the prime factors of a large number: if 65,538 is the number given, we see it is even, hence it admits 2 as a prime factor; the sum of its figures is a multiple of 9, so it admits 3×3 as prime factors and as $(6+5+8)-(5+3)=11$, it is also divisible by 11.

If we put down, as follows, the successive divisions, we see that after division by 2, the number is odd, so 2 is the highest power of 2 that goes into 65,538. After division by 3, the result is still divisible by 3. Then 3,641 is divisible by 11, because $(3+4)-(6+1)=0$ and 331 is prime as we see from our Table in Section 2.

65,538	2
32,769	3
10,923	3
3,641	11
331	331
1	

But if we have no Table by us and we want to find out whether 331 is prime, we need only try to divide it by 13, 17, 19, for if 331 were composite and it admitted 23 as a factor, the other factor would have to be smaller than 19, and we should have already found it, because $19 \times 23 = 417$ which is bigger than 331.

Finding all factors of a number

9. This last observation is very important because it tells us how many trials can be saved in finding whether a given number is prime.

Again, if we know that the square of a prime is just bigger or smaller than the number we are examining, we only need to try all the primes smaller or equal to that one. Here 331 is smaller than $19^2 = 361$. So we only need to try 13, 17 after we have tried 2, 3, 5, 7, 11. If the division is not exact, then the conclusion is that 331 *is* prime.

37

10. Find all the prime factors of

36, 81, 121, 144, 256, 385, 441, 625.

From the prime factors, we can form all the factors. Let us consider the example of 36:

$$36 = 2 \times 2 \times 3 \times 3$$

We therefore have the factors: 1, 2, 2^2 (1)

                      1, 3, 3^2 (2)

and by multiplying the terms of one line by each term of the other we get:

1×1	1×2	1×2^2
1×3	2×3	$2^2 \times 3$
1×3^2	2×3^2	$2^2 \times 3^2$

which, worked out, give the following numbers:

1, 2, 3, 4, 6, 9, 12, 18, 36

Of course, we already know how to find these factors using the tower of rods representing 36.

Had we chosen 144 or 36×4 we would have increased the number of factors, but not the difficulties; the first line (1) would have included 2^3 and 2^4, while the second would have remained the same.

Find all the factors of the numbers set out at the beginning of this section, using the rods or the method above.

We can also discover how many factors a given number has without first working them out and then counting them.

When we were finding all the factors of 36, we multiplied each of the terms of line (1) by each of the terms of line (2) and we found there were nine. But we can arrive at the same answer in a simpler way.

Look at lines (1) and (2) again and you will see that the highest power of each of the two prime factors was 2. If we add *one unit* to each of the highest powers we have found and multiply the numbers so obtained we get, once more, 9.

$$(2+1) \times (2+1) = 9.$$

Let us try this with 144.

$$144 = 2^4 \times 3^2.$$

38

The highest powers of the prime numbers we have found are 4 and 2.

$$(4+1)\times(2+1)=15.$$

This is the same number as is obtained if we make use of the longer method:

1	2	2^2	2^3	2^4
1	3	3^2		

The factors are, thus:

1, 2, 3, 4, 6, 8, 9, 12, 16, 18, 24, 36, 48, 72, 144

Now find the factors of each of the numbers set out at the beginning of this section, but this time use your rods and then the method just described. Check the results to see if they agree.

Exercises and curiosities

11.   If we write a number following the order of the figures, but starting from the left instead of the right, we say we have *reversed* the number. Thus 23 and 32, 134 and 431, 7,293 and 3,927 are the reverse of each other.

Write any number and its reverse and substract one from the other.

You will always find that the result is a multiple of 9.

If your number has an even number of figures and you now *add* the number and its reverse, your result will always be a multiple of 11.

Write any number of three figures. Reverse it and subtract one from the other (putting 0 in the place for the hundreds if it is empty). Reverse the number so obtained, placing it underneath. Then add these last two numbers together and the answer is always 1089.

Square 11, 101, 1001, etc. What numbers do you find?

Take the cube of the same numbers. In Book 7 you will see why it always works out in this way.

Similarly, square 11, 111, 1111 ... up to 9 figures. Again in Book 7 you will find a way of explaining this result.

Find the prime factors of those same numbers (made of the figure 1) up to, say, 111,111.

12. The number 123,456,789, formed of the 9 consecutive figures, has curious properties. If we reverse it and add the two numbers, what do we find?

If we subtract it from its reverse, do we lose any of its figures?

Now consider the number 12,345,679, which omits 8 from the sequence. Repeat what was done with the former number.

Multiply it by 9; what is the answer?

Multiply it by all the multiples of 9 up to 81; what are the answers?

Write it out three times in a line but inserting zero between each complete series. Multiply by 9; what do you find?

13. If, in the formula x^2+x+41, you replace x successively by 0, 1, 2, 3, 4, 5, ... up to 39, prove that you always find a prime number. But if you replace x by 40, the answer is 41×41 (i.e., 41^2) and if you replace x by 41, the answer is 41×43, a composite number.

There is another formula which you can try:

In x^2+x+17, replace x successively by 0, 1, 2, ... 16. Are the results all prime? What happens if x is 16? And what if it is 17?

Remarks on prime numbers

14. When the numbers are small, you usually find they have few properties. When they become larger they are cumbersome and, though they may have many properties, these are difficult to find. You will have discovered that already.

What happens when we are confronted with large numbers is that we no longer ask the same questions as we do with small numbers. For example, nobody will attempt to find

40

whether a number of 257 figures, which you can write in a few moments, is prime or not, unless it is even or ends in a 5 or has for the sum of its figures a multiple of 3, or some easy property such as the one we found for multiples of 11.

The kind of question we ask, when faced with such numbers, is of this type: is there an *infinite* number of numbers having this or that property? If the answer is *no*, then we can attempt to find all those numbers which have that property.

If the answer is *yes*, then we know something important and we also know there is no need to attempt to find the numbers having that property since we cannot find them all.

Here are two examples of that type of question:

(a) Is the number of prime numbers infinite? (That means, are there always primes beyond any number?)

This can be proved, as we shall see later on in this book.

(b) Is there an infinite number of pairs of *consecutive primes* such as 3, 5; 5, 7; 11, 13; 17, 19; . . . ; 101, 103; . . . ?

This question was asked a long time ago, but cannot yet be answered because it is too difficult.

In Part V of this book, you will find more questions of this type.

IV

H.C.F. AND L.C.M.

PART IV

H.C.F. AND L.C.M.

Highest common factor

1. We know how to find the factors of any composite number. If these are arranged in increasing order, and if they are compared with similarly ordered sets of factors of other numbers, we can detect

(a) which are their common factors,

(b) which among these is the largest.

For example, the factors of

36 are *1*, *2*, *3*, *4*, *6*,   9, *12*,   18,   36,
48 are *1*, *2*, *3*, *4*, *6*, 8,   *12*, 16,   24,   48

and among the several italicised common factors, *12* is the largest. It is called the *highest common factor* of 36 and 48, and is usually written as

(36, 48)=12

and read as 'the H.C.F. of 36 and 48 is 12'.

If, instead of two numbers, we take three or four, the process is the same. We find all the factors of each, then the factors that are common and, finally, the largest of these.

This is their H.C.F.

Example: 36, 48, 64,
     64: 1, 2, 4, 8, 16, 32, 64

We have just found the factors of 36 and 48 and need not repeat them. Comparing the three series of factors, we find that 1, 2 and 4 are common and 4 is the highest. It is therefore the H.C.F.

(36, 48, 64)=4

2. Find the H.C.F. of

(a) 34, 51, 119   (d) 35, 75, 105, 145

(b) 37, 111, 666   (e) 57, 91, 42, 108

(c) 72, 54, 108   (f) 41, 43, 47

3. When the H.C.F. of two or more numbers is equal to 1, they are said to be *relatively prime*. Note that when two numbers are prime, they are also prime to each other; but if they are relatively prime, they are not necessarily prime themselves.

For example (4, 9)=1 but neither 4 nor 9 is prime.

Decomposition in prime factors

4. We learnt in Part III how to find the prime factors of any number and, from this, all the factors of that number.

When we want to find the H.C.F. of two numbers, we do not need to find all the factors as we did in Section 1. The prime factors are enough.

Let us form the tower: 'yellow, dark green, tan'. The result of the multiplications is 240.

We see that $240=5\times6\times8=6\times5\times8=8\times5\times6=5\times48=6\times40=8\times30$, so that we have some of the factors of 240. But, in order to have them all, we must replace the dark green by the cross 'red, light green', and the tan by the tower 'red, red, red'. So the new tower is 'red, red, red, light green, yellow', and we have

$$240=2\times2\times2\times2\times3\times5=2^4\times3\times5$$

This is the *decomposition of* 240 *into prime factors*.

Going back to 36 and 48, we know that we can write,

$$36=2^2\times3^2 \quad \text{and} \quad 48=2^4\times3.$$

Since $2^4=2^2\times2^2$ and $3^2=3\times3$, we see that $2^2\times3$ belongs to (i.e., is a factor of) 36 *and* 48, and is also such that if we divide 36 and 48 by it, the results, which are respectively 3 and 4, are prime to each other.

So now we have found another way of looking at the H.C.F. of two or more numbers.

If we divide such numbers by their H.C.F. the quotients obtained are relatively prime.

Test that with the examples worked out in Section 2 or any other group you care to write down.

Repeat the exercise of finding the H.C.F. of the groups of Section 2 using the writing of each number as product of powers or primes.

Lowest common multiples

5. In Book 1, we learned how to make 'trains' of rods of a single color and noticed that we could put two or more trains of different colors side-by-side to form equal lengths. We shall now use these trains for a new purpose.

If we take the white rod as our unit of measurement, a train composed of red rods gives the following sequence of numbers:

2, 4, 6, 8, 10, 12, 14, 16, 18 . . . .

If green rods are used, the sequence will be:

3, 6, 9, 12, 15, 18, 21 . . . .

With pink we get:

4, 8, 12, 16, 20, 24 . . . .

Write down similar sequences for yellow, dark green, black, tan, blue and orange trains.

If we construct two such trains side-by-side, we see that they are equal at certain points. For example, red and green trains are equal when their lengths are 6, 12, 18, 24 . . . .

The shortest equal trains we can make with these rods are equal in length to 6 white rods, the red containing 3 rods and the green 2 rods. If we consider all the multiples of 2 and 3, we see that there is an infinite sequence: 6, 12, 18, 24, 30, 36 . . . . But the *least common multiple* (or L.C.M.) is 6. This is how we denote it: $[2, 3]=6$.

6. With red and yellow trains, doing the same thing, we find

[2, 5]=10, or the L.C.M. of 2 and 5 is 10.

Make trains of rods of one color and put two of different colors side-by-side. Write down the lengths of the trains which can be made with either one of the two types of rods and underline the smallest number common to both trains.

What do you find if the rods are
    yellow and black,
  or yellow and tan,
  or black and blue,
  or pink and yellow,
  or pink and black?

From this, find the L.C.M. of

[5, 7]=   [5, 8]=
[7, 9]=   [4, 5]=
[4, 7]=
Now find [2, 4]=   [4, 6]=   [6, 8]=

7. If instead of only two trains you put three side-by-side (each being of a different color) and you note the first point at which the three are equal you find the common multiple of the three. This is the L.C.M. because it is the smallest length into which the three rods will fit exactly. Thus,

[2, 3, 5]=30, because 30 is a multiple of 2, 3 and 5, and there is no smaller multiple common to these three numbers.

We can see this in another way:

The multiples of 2 are: 2, 4, 6, 8, 10, 12, 14, 16, 18, 20, 22,

24, 26, 28, 30 . . . .
          *

The multiples of 3 are: 3, 6, 9, 12, 15, 18, 21, 24, 27, 30 . . .
                                                              *

47

The multiples of 5 are: 5, 10, 15, 20, 25, 30 . . . .

We have marked the common multiples of 2 and 3 by underlining them once, the common multiples of 2 and 5 by underlining them twice, and those of 3 and 5 by underlining them three times. The multiple common to all three is marked with an asterisk.

If we were to lengthen the three lines so as to reach 60 or 90 or 120 (still using the white rod as our unit of measurement) these would be multiples common to 2, 3 and 5, but 30 is the shortest length at which all three trains are equal and, thus, 30 is their L.C.M.

Work this out similarly with the following rods or lengths of rods or, where the lengths become too great, on paper:

[2, 4, 5] =   [3, 4, 5] =   [3, 5, 7] =
[2, 6, 10]=   [4, 6, 8] =   [6, 8, 10]=
[3, 6, 9] =   [5, 10, 15]=   [14, 15] =

8. Let us now calculate the L.C.M. of two larger numbers, say, 36 and 48.

We can imagine constructing two trains side-by-side using carriages of the length of 36 for one and 48 for the other. If we went on adding such carriages we should find the first point at which the trains were equal and this would give us the L.C.M. of those numbers.

To do this with rods would be complicated and a great waste of time, but we can see a simpler way. $36=2^2\times3^2$ and $48=2^4\times3$. We could make the length we are looking for with any of the rods which represent the common factors we have found, so we could use red, green or pink rods. We could also make it with carriages equal in length to the highest common factor, which we already know how to find: $(36, 48)=12$.

But $36=12\times3$ and $48=12\times4$, so that 3 and 4 are the *non-common factors*.

48

To obtain the L.C.M. of 36 and 48 we must multiply their H.C.F. by the L.C.M. of their non-common factors. The L.C.M. of 3 and 4 is 12, so $12 \times 12 = 144$ is the L.C.M.

The trains we were constructing in our imagination would be made up of 3 carriages 48 in length and 4 carriages 36 in length, the length of each being 144. We could not make any shorter trains with these carriages which would be equal.

If we take 35 and 42 as our numbers we find their L.C.M. in the same way but shall write what we do more briefly:

$$(35, 42) = 7$$
$$35 = 7 \times 5 \quad \text{and} \quad 42 = 7 \times 6$$

So $[35, 42] = 7 \times [5, 6] = 7 \times 30 = 210$

If our two numbers were 19 and 37, since $(19, 37) = 1$ i.e. are relatively prime, $[19, 37] = 19 \times 37 = 703$.

For three numbers 8, 15, 24, since $8 = 2^3$, $15 = 3 \times 5$ and $24 = 2^3 \times 3$, their H.C.F. is 1. Their L.C.M. is found by writing down each of their prime factors once in increasing order, but with each prime factor raised to the highest power at which it appears in any of the three decompositions we have found. So,

$$[8, 15, 24] = 2^3 \times 3 \times 5 = 120.$$

When the H.C.F. is other than 1, we divide each number by the H.C.F., and then multiply the L.C.M. of the quotients so obtained by that H.C.F. Here is an example worked out:

To find $[36, 48, 144]$

$$(36, 48, 144) = 12 \qquad \frac{36}{12} = 3 \qquad \frac{48}{12} = 4 \qquad \frac{144}{12} = 12$$

$$[3, 4, 12] = 12$$

Hence, $[36, 48, 144] = (36, 48, 144) \times [3, 4, 12]$
$$= 12 \times 12$$
$$= 144$$

Of course we might have guessed the answer because we have just worked with 36 and 48, and $144 = 4 \times 36 = 3 \times 48$.

Test this on the following examples:

[9, 24]　　　　　　[9, 15, 24]　　　　　[14, 35]

[10, 15, 25]　　　　[22, 33, 66]　　　　[34, 51, 85]

V

SQUARES, CUBES AND
SQUARE ROOTS

PART V

SQUARES, CUBES AND SQUARE ROOTS

Squares

1. We already know that with the rods we can make squares and cubes (see Books 2 and 6). Find the answers to:

1^2	2^2	3^2	4^2	5^2
6^2	7^2	8^2	9^2	10^2
1^3	2^3	3^3	4^3	5^3
6^3	7^3	8^3	9^3	10^3

Square of a sum

2. We can make lengths that are bigger than the orange rod by putting a certain number of rods end-to-end. We are now going to see how we can form the squares and the cubes on these lengths.

In Book 2, we have already seen in the calculation of 29^2 and 31^2, that the square of a sum or a difference can be found with the rods.

Let us start with the length $11=5+6$, made of a yellow and a dark green rod. The square is easily made by taking eleven yellow rods side-by-side and eleven dark green, and placing the two rectangles end-to-end. If from that square we remove, at two opposite corners, a yellow square made of 5 rods, and a dark green square made of 6 rods, we are left with two rectangles, one made of 5 dark green rods and the other of 6 yellow rods. They can be superimposed, so they are equal.

Hence the original square, 11^2 or $(5+6)^2$, is made of two squares 5^2 and 6^2, and of twice a rectangle 5×6. This can be written,

$$(5+6)^2 = 5^2 + 6^2 + 2 \times 5 \times 6.$$

Repeat this with

$11 = 4 + 7$	or $11 = 3 + 8$	or $11 = 2 + 9$
$13 = 5 + 8$	or $13 = 7 + 6$	or $13 = 10 + 3$

3. Of course, you can see that the large square always *contains* the two smaller ones, so that we can say: The square on the sum of two rods is *greater* than the sum of the squares on each of the rods. Or, in other words, the square of the sum is larger than the sum of the squares. We also see by how much it is bigger, for the square on the sum is equal to the sum of the squares plus twice the rectangle on the two given lengths.

Let us write all the examples we can form with the rods, no longer using figures, but the initial of the color of the rods.

Thus $y + r$ is for yellow and red end-to-end,

$(y+r)^2$ is for the square on the yellow and the red end-to-end,

y^2 is for the square on the yellow, and r^2 for the square on the red, and, $y \times r$ is for the rectangle yellow, red.

We can then write

$$(y+r)^2 = y^2 + r^2 + 2 \times y \times r$$
$$\text{or } (r+y)^2 = r^2 + y^2 + 2 \times y \times r$$

Give similarly what you would find if you formed the squares on the following lengths, $b+y$; $p+d$; and $o+w$.

Can you test whether the following writings are correct?

$$(o+r)^2 = o^2 + r^2 + 2 \times o \times r$$
$$(t+g)^2 = t^2 + g^2 + 2 \times t \times g$$
$$(b+d)^2 = d^2 + b^2 + 2 \times d \times b$$
$$(t+b)^2 = t^2 + 2 \times t \times b + b^2$$
$$(r+g)^2 = g^2 + 2 \times r \times g + r^2$$

4. Must we limit ourselves to two rods end-to-end?

Put $r+g+p$ and make the square on it and compare the result to $r^2+g^2+p^2$. Can you say again that the square on the sum is *greater* than the sum of the squares? Can you find by how much?

We can see that the comparison could be done like this:

Since $r+g=y$ we can write

$$(r+g+p)^2=(y+p)^2$$

and we already know that $(y+p)^2=y^2+p^2+2\times y\times p$.

Or

$$(y+p)^2=(r+g+p)^2=(r+g)^2+p^2+2\times(r+g)\times p$$

But $(r+g)^2=r^2+g^2+2\times r\times g$. So that

$$(g+r+p)^2=r^2+g^2+2\times r\times g+p^2+2\times r\times p+2\times g\times p.$$

Or, rearranging the terms,

$$(r+g+p)^2=r^2+g^2+p^2+2\times r\times g+2\times r\times p+2\times g\times p.$$

So the sum of squares is smaller than the square on the sum, and the difference is made of twice the rectangles that can be made with the rods taken two-by-two.

Find,

$$(p+g+d)^2$$
$$(o+b+p)^2$$
$$(g+t+y)^2$$
$$(r+p+y)^2$$

Can you test whether the following are correct?

$$(r+d+b)^2=r^2+2\times d\times b+2\times d\times r+b^2+d^2+2\times b\times r.$$
$$b^2+o^2+t^2+2\times t\times o+2\times t\times b+2\times b\times o=(o+t+b)^2$$

5. There is no difficulty in extending the formation of a large square on four or more rods and comparing the result to the sum of the squares. The difference is always made of as many pairs of equal rectangles as there are ways of pairing the rods.

Check

$$(r+g+y+b)^2=r^2+g^2+y^2+b^2+2\times r\times g+2\times r\times y+2\times r$$
$$\times b+2\times g\times y+2\times b\times g+2\times y\times b.$$

and find

$$(g+p+y+d)^2$$
$$(p+o+b+t)^2$$
$$(g+d+b+o)^2$$

Formation of a table of squares

6. All this will help us in the making of the *Table of Squares* up to 100. We already know that

$1^2=1$	$3^2=9$	$5^2=25$	$7^2=49$	$9^2=81$
$2^2=4$	$4^2=16$	$6^2=36$	$8^2=64$	$10^2=100$

But we can find out at once that

$20^2=400$	$40^2=1600$	$60^2=3600$	$80^2=6400$
$30^2=900$	$50^2=2500$	$70^2=4900$	$90^2=8100$

and $100^2=10,000$.

In order to find the squares of the other numbers, we can use what we have learnt in sections 3 to 5.

$$11^2=(10+1)^2=10^2+1^2+2\times10\times1=\ 100+\ 1+\ \ 20=\ 121$$
$$12^2=(11+1)^2=11^2+1^2+2\times11\times1=\ 121+\ 1+\ \ 22=\ 144$$
$$13^2=(10+3)^2=10^2+3^2+2\times10\times3=\ 100+\ 9+\ \ 60=\ 169$$
$$24^2=(20+4)^2=20^2+4^2+2\times20\times4=\ 400+16+\ 160=\ 576$$
$$37^2=(30+7)^2=30^2+7^2+2\times30\times7=\ 900+49+\ 420=1369$$
$$79^2=(70+9)^2=70^2+9^2+2\times70\times9=4900+81+1260=6241$$
$$85^2=(80+5)^2=80^2+5^2+2\times80\times5=6400+25+\ 800=7225$$

Of course, there are many ways of obtaining a result and we have chosen one at random. You are free to choose the one you find simplest so long as your result, when checked, is correct.

Complete the following:

TABLE OF SQUARES

No.	Square	No.	Square	No.	Square	No.	Square	No.	Square
1		21		41		61		81	
2		22		42		62		82	
3		23		43		63		83	
4		24		44		64		84	
5		25		45		65		85	
6		26		46		66		86	
7		27		47		67		87	
8		28		48		68		88	
9		29		49		69		89	
10		30		50		70		90	
11		31		51		71		91	
12		32		52		72		92	
13		33		53		73		93	
14		34		54		74		94	
15		35		55		75		95	
16		36		56		76		96	
17		37		57		77		97	
18		38		58		78		98	
19		39		59		79		99	
20		40		60		80		100	

7. Let us examine in particular the squares of numbers that end with 5. We notice that they all end with 25 and, if we look at the number represented by the figure or figures to the left of 25 we find that it is the product of the number represented by the figure or figures to the left of the 5 (in the number to be squared) and that same number plus one.

Thus $35^2=1225$ and $12=3\times4$; $75^2=5625$ and $56=7\times8$.

To make sure that this is always so, let us write a number ending in 5 as: $10\times a+5$, where a is 1 or 2 or 3 or 4 or any other whole number.

$10a+5$ gives all numbers ending in 5 if we replace a by all successive integers.

If $a=1$, $10\times1+5=15$
$a=2$, $10\times2+5=25$

56

$$a=3, \ 10 \times 3 + 5 = 35$$
$$a=12, \ 10 \times 12 + 5 = 125$$
$$a=17, \ 10 \times 17 + 5 = 175 \text{ etc:}$$
$$\text{But } (10 \times a + 5)^2 = (10 \times a)^2 + 5^2 + 2 \times (10 \times a) \times 5$$
$$= 100 \times a^2 + 25 + 20 \times a \times 5$$
$$= 100 \times a^2 + 100 \times a + 25$$
$$\text{or,} \quad 100 \times a(a+1) + 25$$

$a+1$ *is* the number that is bigger than a by one unit. So the number of hundreds of the square is found to be a multiplied by $a+1$, as we noticed.

8. It is clear that we can multiply any number by itself and form a Table of Squares that goes as far as we care to pursue it but, here, we have discovered how to save energy and time by using the results already obtained in the Table to continue it. A Table once made can be used to save time in calculations or even to avoid calculating squares altogether.

Square of a difference

9. The square and the cube of a sum can actually be performed with rods, but the square and the cube of a difference require us to perform some of the manipulations mentally.

The square of 8 is 64 and can be made with 8 tan rods. If we write $8 = 9 - 1$, we can use the square made of blue rods and the white rod. It is clear that the blue square does not equal the tan square plus two blue rods, but the blue square plus one white rod does. You can see this if you place the two squares near each other and add the two blue rods to the tan and one white rod to the blue. If we now remove (mentally) from the blue-square-plus-one-white two rectangles each equal to the area of the two blue rods, we are left with an area equal to that of the tan square. Or we can, if we wish, instead, place the tan square on the blue-square-plus-one and see that we are left with the area of two blue rods showing. So we can write,

$$9^2 + 1 - 2 \times 9 \times 1 = 8^2 = (9-1)^2$$

Similarly, $8=10-2$. If from the orange square to which we have added a red square we remove an area equal to twice the rectangle whose dimensions are the orange and red, we are left with an area equal to the tan square. Or,

$$10^2+2^2-2\times10\times2=8^2=(10-2)^2$$

We can put this in a more general way. If we make the tan square first and border it with two orange rods on one side, completing the square with two more tan rods, we have constructed the area of an orange square. If we then add a red square at one end of the orange pair we can see that the tan square is equal to the orange plus red squares less two rectangles whose dimensions are the orange and the red. If we superimpose two pairs of orange rods this becomes obvious. So we can say: the tan square is equal to the orange and red squares added to each other with two rectangles, each made of two orange rods side-by-side, subtracted. Or,

$$t^2=(o-r)^2=o^2+r^2-2\times o\times r$$

If we do not add the smaller square we cannot cut off two equal rectangles.

This relationship can serve for squares of any dimensions and can be used for finding squares such as 79^2 more easily, for, $79^2=(80-1)^2=80^2+1^2-2\times80\times1=6401-160=6241$ which is easier than multiplying 79 by 79 in the usual way.

Formation of a table of cubes

10. *A Table of Cubes* can be made as easily, as soon as we have learned how to form cubes.

With the rods, we can form the cubes of 1, 2, 3, 4 and, if we have enough rods, of 5, 6, 7, 8, 9, 10. We already know that we need 1, 4, 9, 16, 25, 36, 49, 64, 81, 100 of each of the respective rods to produce these cubes if we use rods of only one color. But we can, if we wish, form the cube of 5 using red and light green rods. Let us do it.

If we put one layer of 25 red rods, each standing on end, with 25 light green ones standing upright on top we form a cube congruent with the cube we would have if we had used only yellow rods. If we remove 9 light green rods

58

from one corner and put it beside the original cube, we have a light green cube, and we can see that *if* we removed 4 red rods from the opposite corner—which we cannot do without making the cube collapse—we would get a red cube as well. So, we can at once say that the cube of the sum is bigger than the sum of the cubes, and we can see by how much it is bigger.

Thinking as we did when dealing with the squares, we can look for the volume of what is left of the original cube as volumes of rectangular prisms. One is made of six red rods and, on top, there are six light green ones. If we remove them and put them aside, we see that there are two prisms on the opposite side, one equal to the red prism and one to the light green prism, and we are left with a red prism under the light green cube and a light green prism above the red cube. So, in all, we have in the 'yellow' cube: one light green cube, one red cube, 3 prisms with base 2×3 and height 2, and 3 prisms with base 3×2 and height 3. As the dimensions 2 and 3 are represented by the red and light green we can write:

$$(g+r)^3 = g^3 + r^3 + 3 \times g \times r^2 + 3 \times r \times g^2.$$

Make the dark green cube in the same way, using red and pink rods and check whether the following writing is correct:

$$(r+p)^3 = r^3 + p^3 + 3 \times r^2 \times p + 3 \times r \times p^2.$$

Find the cube of 6, 7, 8, 9, 10 using this method after you have calculated 1^3, 2^3, 3^3, 4^3, 5^3.

11. You see that you can make use of your Table of Squares at once to form a Table of Cubes. So, in the easy example:

$$13^3 = (10+3)^3 = 10^3 + 3^3 + 3 \times 10^2 \times 3 + 3 \times 10 \times 3^2$$
$$= 1000 + 27 + 3 \times 100 \times 3 + 3 \times 10 \times 9$$
$$= 1000 + 27 + 900 + 270$$
$$13^3 = 2197.$$

For 20^3, 30^3, . . . 100^3, we can note that

$$20 = 2 \times 10, \quad 20^3 = 2^3 \times 10^3 = 8000.$$

Find these answers and use them to complete the following:

TABLE OF CUBES

No.	Cube	No.	Cube	No.	Cube	No.	Cube	No.	Cube
1		21		41		61		81	
2		22		42		62		82	
3		23		43		63		83	
4		24		44		64		84	
5		25		45		65		85	
6		26		46		66		86	
7		27		47		67		87	
8		28		48		68		88	
9		29		49		69		89	
10		30		50		70		90	
11		31		51		71		91	
12		32		52		72		92	
13		33		53		73		93	
14		34		54		74		94	
15		35		55		75		95	
16		36		56		76		96	
17		37		57		77		97	
18		38		58		78		98	
19		39		59		79		99	
20		40		60		80		100	

We already know how to estimate which is the better of two procedures. Let us see whether we are saving time and energy in the task of making the Table.

First, for the example of 79^3, we can see that $79=70+9$, or $60+19$, or $50+29$, or $75+4$ etc.

$79^3=(75+4)^3=75^3+4^3+3\times75^2\times4+3\times75\times4^2$.

75^3 has been calculated by the time we reach 79^3, and is 421,875.

4^3 is 64, 75^2 is in the previous Table: 5,625, $4^2=16$.

So $79^3=421,875+64+67,500+3,600=493,039$.

This has taken more time than working out $(79\times79)\times79$, which contains four multiplications and two additions. The calculation we have just done has four multiplications and three additions.

If we know how to calculate $(80-1)^3$, all the calculations in this case would have been much easier.

Cube of a difference

12. We can use the rods to show very easily the square and the cube of any sum. In order to get the square and the cube of a difference, we must first do a few manipulations mentally.

The square of 8 is 64, and can be formed of 8 tan rods. But it can also be written $8^2 = (9-1)^2$, using the square of the blue rod minus the white rod. We can see that the blue square is not equal to the tan square plus two blue rods, but that if we remove one blue rod from the blue square and add a white rod to the rectangle that remains, and then take away from this once again the area of one blue rod, the area remaining is equal to the tan square. We can therefore say that

$$(9-1)^2 = 9^2 + 1^2 - 2 \times 9 \times 1$$

The same thing happens for $8^2 = (10-2)^2$. The tan square is equal to what is left of the orange square when, after adding a red square, we remove two rectangles, whose dimensions are the orange and the red.

From this we have

$$10^2 + 2^2 - 2 \times 10 \times 2 = (10-2)^2 = 8^2$$

If we make a tan square and place on two of its sides two rectangles each made of two orange rods we obtain a new square with another square adjacent to it. We can say that he tan square is equal to the sum of an orange square and a ⁺ed square, less the area of the two rectangles each formed of rwo orange rods placed side by side.

Thus:

$$t^2 = (o-r)^2 = o^2 + r^2 - 2 \times o \times r$$

It is only by adding the smaller red square to the orange square that we are able to obtain two equal rectangles beside the tan square.

This relationship is valid for all squares of all sizes. It can be used to calculate the squares of numbers such as 79 more easily. We can write

$$79^2 = (80-1)^2 = 80^2 + 1^2 - 2 \times 80 \times 1$$

If we know that

$$8^2=64 \text{ and } 80^2=6{,}400, \text{ we have}$$
$$79^2=6{,}401-160$$
or $\quad 79^2=6{,}241$

which is an easier calculation than multiplying 79×79 in the usual way.

Let us now study the cube of a difference and take, as an example for this, $y-r$, the difference between a yellow rod and a red rod.

Form three prisms of $y^2 \times r$, that is, three prisms each of which is formed of two yellow squares placed on top of the other. Place one of these prisms in a horizontal position, flat on the table. The remaining prisms are to stand vertically along two contiguous sides of the top of the flat prism. Note that one of these prisms extends beyond the square surface of the horizontal prism, with a portion of its volume overhanging the prism on which it stands, as is seen in the figure.

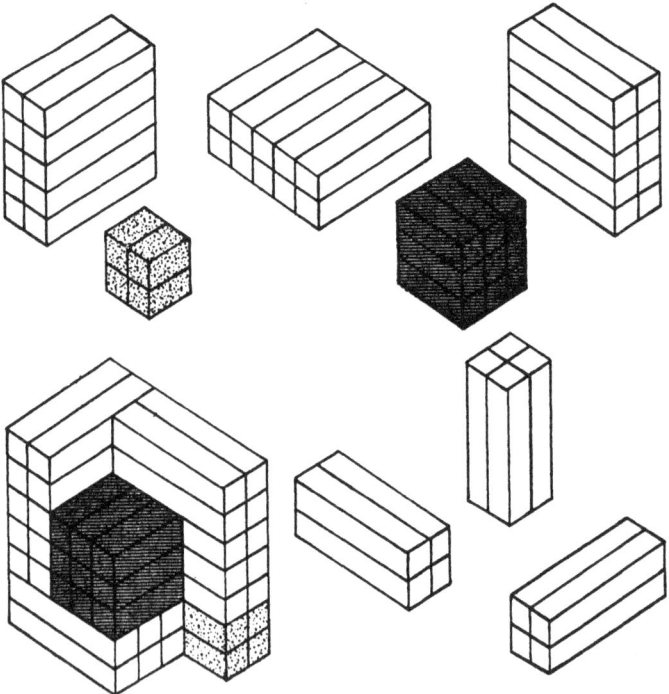

Place a light green cube on what remains of the top of the horizontal prism, in the corner formed by the two vertical prisms. Finally, place a red cube under the unsupported part of the yellow prism. The body we have just made is, then, the sum of the first three prisms, $y^2 \times r$, plus the light green cube and the red cube:

$$g^3 + 3\,(y^2 \times r) + r^3$$

But this same body can also be made by using a cube of dimensions equal to the length of the yellow rod, and three prisms of the size $r^2 \times y$. Two of the prisms ($r^2 y$) can be found in a horizontal position at the top of the body we are studying and the third, which is vertical, is that made by the red cube and two thirds of the part of the yellow prism which it supports. The volume of this body, then, can also be expressed as

$$y^3 + 3 \times r^2 \times y$$

We therefore have two bodies equal in volume but one made of

$$g^3 + 3y^2 r + r^3,$$

and the other of $\qquad y^3 + 3r^2 y$

If we subtract from both the volumes represented by $3y^2 r + r^3$, the first becomes the light green cube g^3 while the second becomes $y^3 + 3r^2 y - 3y^2 r - r^3$.

Hence

$$g^3 = y^3 - 3y^2 r + 3r^2 y - r^3 = (y - r)^3$$

re-writing the various terms so as to have a decreasing order of the powers of y and an increasing order of the powers of r.

Repeat the construction above using other colored rods or cubes and prisms and test the generality of the relationship above.

This we shall write using any letters, for example: x and u

$$(x - u)^3 = x^3 - 3x^2 u + 3\,x\,u^2 - u^3$$

Square roots

13. Looking at our Table of Squares, it is clear that only a few numbers between 1 and 10,000 have a corresponding

63

number in the other column; in fact, there are only one hundred. Between 80 and 99 for example, only 81 has a corresponding number.

When we start with the numbers, we find the squares in the next column. If we now look at the numbers in the column of squares first, those in the other column are called their *square root*. Thus, 9 is the square root of 81 and, in notation, we have

$$9^2 = 81 \quad \text{or} \quad 9 = \sqrt{81}$$

If we want to make a *Table of Square Roots*, only those hundred numbers in the columns of squares can give us a whole number as their square root and, if we ask what the square root of 80 or 90 is, we cannot give an answer that is a whole number. So we must understand more exactly what we mean by squares and square roots.

With the rods, the squares we make are all multiples of one square centimetre, i.e., the face of the white rod. But if we draw a square, it is most unlikely that its side will be an exact number of centimetres long. If we draw a square on 3·5 cms, we do not obtain a whole number of square centimetres. In that case, we get 12·25 sq. cms.

We know how to multiply decimals and so we could, if we wanted, obtain Tables of Squares much longer than the one we have. For instance, if we square 1·1, 1·2, 1·3, ... 1·9, 2, 2·1, 2·2, ... 2·9, 3, ... we form a Table in which the squares have two figures on the right of the decimal point. But we do not obtain all the numbers with two figures. For instance, between $(1·1)^2 = 1·21$ and $(1·2)^2 = 1·44$, we have no other number in our Table.

We could take numbers with two or more figures on the right of the decimal point and square them. But even then there would be many decimal numbers that would not appear in our Table.

Approximations

14. This will help us to grasp an important idea.

The rods are concerned with whole numbers or a few fractions, and the numbers we can think of are much more numerous than we could define using the rods. In Book 4, we saw that some fractions have decimal forms that never end, some being recurrent and some being periodic decimals. If we think of writing a decimal *at random*, with an infinite number of decimals, it will not be periodic; and, as we shall see in Book 7, there is no fraction equal to it.

So we have whole numbers, fractions, and infinite decimal fractions, periodic or non-periodic, to consider. We can operate on all of them.

Decimals with an infinite number of figures on the right of the decimal point can only be thought of; they cannot actually be written. What we can actually do is always restricted to finite expressions. The decimals which cannot be written, but which we consider in a form that serves our purpose, are called *approximations*, and we must learn to work with them since they are the only ones we meet in practice.

If we ask what the square root of 80 is, we know it is not 9 but must be a number *very near* 9. The feeling of 'very near' must be expressed more precisely if we are to share what we mean with others, and this is the problem of approximation that is presented by most calculations arising out of everyday life where some degree of precision matters.

The range of possible numbers that could appear in any column of a Table of Squares is immense and we can see that relatively few of them can actually be included. We must therefore find ways of expressing exactly how near the square root of a number is to a number included in our Table.

If the Table contains only the squares of whole numbers, the square root of 79 or of 80 will be seen to be greater than 8, so 8 will be called the square root of these numbers *to the nearest unit*. Though 8 is too small, 9 is too big. We may, of course, prefer to say that 9 is the nearest whole number whose square is 'near' 79 or 80.

To distinguish the two cases we shall call 9 the square root *by excess*, and 8 the square root *by defect*, since the former is greater, and the latter smaller, than the actual number for which we are looking.

If our Table contained the squares of decimal numbers with one figure on the right of the decimal point, we would have a better choice for the square root of 80. We can see that 79·21, the square of 8·9, is smaller than 80, whereas 81, the square of 9, is bigger. So 8·9 is the square root by defect of 80 to the nearest tenth in that Table.

If the Table contained the squares of numbers with two figures to the right of the decimal point, we could find a number that is bigger than 8·9 whose square is nearer to 80 than to 81. 79·9236 is the square of 8·94, whereas 80·1025 is the square of 8·95, so 8·94 is the square root by defect of 80, to the nearest hundredth, i.e. if we add to 8·94 one hundredth (or ·01) the square of the answer is bigger than 80.

All that is required, then, is that we have Tables of Squares containing decimal numbers with as many places to the right of the decimal point as we need for our purpose. But we may want to find the square root of a given number to a given number of decimal places. Can we do this? The answer is, yes, and we shall consider this here also.

Calculating square roots

15. The procedure by which the square root of a given number is obtained can be made clear by diagrams. If we represent the given number by a square we see that what we are looking for is represented by the length of the side of that square.

We can find out the number of digits contained in the answer if we write down the given number and divide it from the right in blocks of two figures. If there is only one figure on the left we treat this, nonetheless, as a block. The number of blocks gives us the number of digits in the square root.

Take 63,742 and divide it in this way:

$$6'37'42$$

There are three blocks, so we shall look for a number that has three digits.

We can at once find the square root of this number to the nearest hundred, for $300 \times 300 > 63,742$ and $200 \times 200 < 63,742$ So we know that the square root will be a three-figure number beginning with 2.

Our task is to find the remaining two digits and the procedure by which we shall carry out that task will be easy to

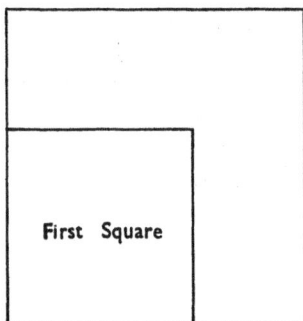

understand if we study the diagrams. These are not intended to reproduce by their proportions the actual numbers with which we are working—or, indeed, any particular numbers. They simply illustrate the principles involved in what we are doing. If we wish we can draw a square and use our rods to reproduce the stages of the calculation.

(a)

First Square

We have placed a square inside the original square with a base that represents 200. Its area represents 40,000 whereas that of the original square represents 63,742. The difference is indicated by the L-shaped portion left over.

We must enlarge this first square until its base represents the square root of the given number to the nearest unit, and we must do this in two stages. The first will give us the square root to the nearest ten and the second to the nearest unit. Unless this proves to be the exact square root of the given number we shall expect still to find an L-shaped remainder. The next diagram illustrates these stages:

(b)

First Square

1 | 2 | R

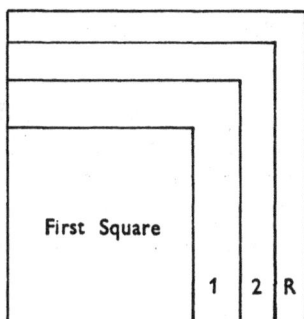

Here we see the first square with its first and second extensions. We also see the remainder, now much reduced. It borders the largest square we have been able to place inside the original square and the base of this largest inside square represents the square root by defect to the nearest unit.

To determine the second (i.e. ten) figure of the square root we must find the length represented by the extension marked 1, and our last diagram shows what we must do to achieve this:

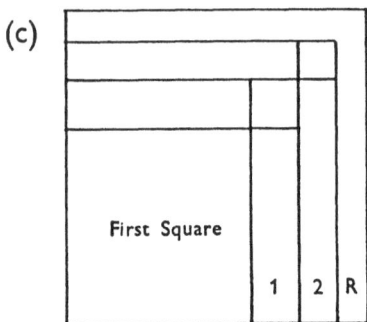

(c)

First Square

1 | 2 | R

The first extension is seen to be made up of two rectangles each bounded by the side of the first square (the length of which we already know) and by a side the length of which we wish to find. A small square completes the extension and each of the sides of this is also the length we wish to find.

So, the area of the first extension is $200 \times 2 \times$ (the number to be found) $+$ that number squared. The extended base will be $200+$ that number (not squared).

When we have found that number we shall have the information we need to deal with the second extension and shall use the same procedure to do it.

If the diagrams, and what they have shown us, are kept in mind we shall easily follow the arithmetical procedure that is set out and explained below, which we shall see resembles long division.

Examples

a) *Find the square root of* 63,742 *to the nearest unit.*

```
      2  5  2
     ‾6'37'42
  45 | 2 37
 502 |  12 42
```

Working : (i) Divide the number into blocks and find the nearest square root by defect of the number in the first block. This we have already found to be 2, so we insert it as the first digit of the answer and subtract 2^2 from 6, leaving 2. Bring down the next block, which gives 237.

(ii) Carry 2 *doubled* out to the left (remembering that there were *two* rectangles). Now find a number which, when placed to the right of the 4, will give a number which, multiplied by the number so found, will be as nearly as possible equal to (though not greater than) 237. By trial and error we find that 5 is the number required, for $45 \times 5 = 225$. This subtracted from 237 leaves 12 which we put down and bring down the last block. The number we have just found, 5, is inserted as the second digit of the answer and is also carried out to the right.

(iii) Just as we doubled 2, this 5 in 25 must be doubled too. This gives us 50 on the left and, once more, we find a number which, placed to the right of 50 will produce a number which, multiplied by the number so found will be as nearly as possible equal to (though not greater than) 1242. We find that the number we need is 2, and this gives us the final digit of the answer.

(iv) If we wish to know the remainder, we subtract 2×502 from 1242 and find 238.

$$\text{So, } \sqrt{63{,}742} = 252 \ r \ 238$$

Here are a few examples to analyse. They introduce some new ideas and the working is set out more fully but, if you remember the diagrams, you will see what is being done and why it is done in that way.

b) *Find the square root of :*

	347,281		4,073,423		11,111,111
	5 8 9		2 0 1 8		3 3 3 3
	34'72'81		4'07'34'23		11'11'11'11
	25		4		9
108	972	401	00734	63	211
×8	864	×1	401	×3	189
1169	10881	4028	33323	663	2211
×9	10521	×8	32224	×3	1989
	0360		1099	6663	22211
				×3	19989
					2222

$\sqrt{347{,}281} = 589$ $\sqrt{4{,}073{,}423} = 2018$ $\sqrt{11{,}111{,}111} = 3333$

r 360   *r* 1099   *r* 2222

Square each of the numbers obtained in the answers after adding 1:

$590^2 = 348,100$ which is greater than 347,281

$2019^2 = 4,076,361$ which is greater than 4,073,423

$3334^2 = 11,115,556$ which is greater than 11,111,111.

We see that the procedure yields the square root by defect to the nearest unit.

Write down large numbers at random. Extract their square roots to the nearest unit and find the remainder. Square your answer, add the remainder, and see if it agrees with the number with which you started.

Write down numbers of three or four digits. Square them, add a 'remainder' at random, and then extract the square root of the resultant number and find the remainder. Does this agree with the numbers with which you started?

Exchange such problems with your neighbor and check your results with each other.

Square roots to a certain approximation

16. If we wanted to go to the nearest tenth of a unit, we would need to add two zeros on the right of the last remainder and continue the procedure. In the following example, it is clear that, for each digit of the square root, we need to place two zeros on the right of the decimal point of the given number.

Find to 3 decimal places the square root of 80. We start by writing 80·000000 with 6 zeros on the right and the working set out below shows that we then proceed as before:

```
                    8 · 9  4  4
                 ─────────────────
                 80·00′00′00
                 64
                 ──────
169              1600
×9               1521
                 ──────
1784             7900
×4               7136
                 ──────
17884            76400
×4               71536
                 ──────
                 04864
√80̄=8·944        r 0·004864
```

If we squared 8·945 we would find 80·023025 which is larger than 80. So 8·944 is the square root of 80 to the nearest $\frac{1}{1000}$ of a unit.

17. In this study of square roots, we have met numbers and decimal numbers that are not the *exact* value that can, in other cases, be found. Our Table of Squares has shown us that, more often than not, the square root of a given number will *not* be exact. We are usually asked, in this type of question, to find a number that *approaches* the answer to a certain value (say, the nearest unit or the nearest $\frac{1}{10}$, $\frac{1}{100}$ . . . of the unit) rather than the exact number. This principle of approximation will become very important in your later studies, and that is why it is so useful for you to understand it in the field of square roots.

VI

THE SET OF INTEGERS

Part VI

THE SET OF INTEGERS

Infinite sets

1. We already know we can get larger and larger numbers by different methods, such as addition of the unit to itself, or addition of a constant number, or multiplication by 2, or exponentiation. All this must have given us the knowledge that the set of whole numbers, or integers (I), is endless or *infinite*.

Of course, we find few opportunities of actually studying very large numbers, and our mind is a blank when we consider numbers that are bigger than a few thousand millions. For instance, all we know of numbers of 50 figures is how to name them. There are no properties of such numbers that we can grasp, because they lie outside the range of experience and imagination.

Equivalent sets

2. In order to study (I) as such, new ideas must be found, and we shall now introduce an idea which can be easily understood if we approach it as follows.

Let us imagine that two people who cannot yet count want to know which of them has more rods (or any other objects). They can agree to put down one object at a time each, and go on doing this until one or both are left with none. If they finish together, although they do not know how many each has, each knows he has as many as the other. If one finishes first, they then know which of them has more and which has less.

This method of matching elements of sets is called the *one-one correspondence*. In notation, if $\{a, b, c, \ldots m\}$ is a set S and $\{a', b', c', \ldots n'\}$ is another S', we can write:

$$a \quad b \quad c \ldots m$$
$$\updownarrow \quad \updownarrow \quad \updownarrow \qquad \updownarrow$$
$$a' \quad b' \quad c' \ldots m' \quad n'$$

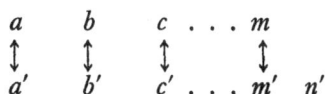

the double arrow indicating that a is matched with a' and a' with a, b with b' and b' with b, etc. If m finishes matched with m', and other elements in S' are left unmatched, we say that S' *contains more elements than* S, or S contains fewer than S'.

If in this one-one correspondence there is always one element of S to go with each one of S' and one of S' to go with each one of S then S and S' are said to be *equivalent* or *equipotent* or to have the *same cardinal number*.

In notation, $S \sim S'$. If $S \sim S'$ then also $S' \sim S$.

Is it true that (1) $S \sim S$ always?

(2) if $S \sim S'$ and $S' \sim S''$ then $S \sim S''$?

Partition of the set of integers

3. If a set S has n elements (say 1,000,000), can you say it is equivalent to S' having m elements (say 900,000)?

To find the answer, we can write down the elements of S one after another and put, beside each, one element of S'. When we actually do it, we see that 100,000 elements of S have no counterparts.

So it is true that if each of two sets has a *finite* number of elements, they can only be equivalent if they have the *same* number of elements.

But with the set (I) we can see a difference. If we separate in (I) the even and the odd numbers, it is clear that to every even there corresponds an odd number, and to every odd there corresponds an even. In notation,

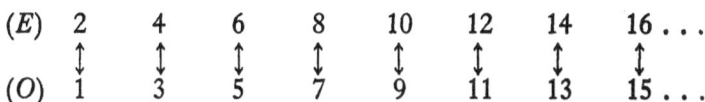

(E)	2	4	6	8	10	12	14	16 ...
	$\updownarrow$	$\updownarrow$	$\updownarrow$	$\updownarrow$	$\updownarrow$	$\updownarrow$	$\updownarrow$	$\updownarrow$
(O)	1	3	5	7	9	11	13	15 ...

It is most important to understand that the actual correspondence continues, and that only time prevents us from writing more and more, always finding that the correspondence is maintained. The dots in the two rows indicate the sequences that are not written down, and imply that what is shown is not terminated at the last number written.

(E) and (O) are equivalent. But they have as many numbers as we want or, as we have said, these sets are endless or *infinite*.

We may have felt that this equivalence was to be expected since, after each odd number, there is an even number, and conversely. There could not be more even than odd, nor more odd than even, numbers.

Equivalence of (I) and one of its parts

4. Now, could we say that there are as many integers as there are even numbers? Or can we find a one-one correspondence between (I) and (E)?

The answer is in the following diagram:

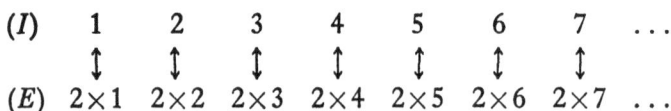

(I)	1	2	3	4	5	6	7	...
	↕	↕	↕	↕	↕	↕	↕	
(E)	2×1	2×2	2×3	2×4	2×5	2×6	2×7	...

This reads: Since the successive even numbers in (E) are each twice the corresponding number in sequence (I), there is for each element of (I) an element in (E) and vice versa. That is to say, (I) and (E) are equivalent.

This seems an extraordinary result—but only if we forget the new characteristic we have found in (I), (E) and (O), namely that each has an endless number of elements. When, in No. 2, we were comparing m and n elements we were basing our procedure upon the knowledge that we must come to the end of S and S'; but now there is no end. The relationship between the elements of (I) and (E) is not altered because we take more and more of their elements for comparison, and it is this relationship, which we can find when we compare *some* elements, that is the link between the two sets. The fact that (E) is only a *part* of (I) opens up new realms of thought instead of disturbing us.

If a set can be put into a one-one correspondence with one of its *true parts*, it is a sign that it cannot be exhausted, or that it is *infinite*. We shall, from now on, always use that term in respect of a set when it can be shown that it can be put into this correspondence with one of its parts.

Other partitions of (I)

5. We are already familiar with

$(I):$ 1, 2, 3, 4, 5 . . .
$(E):$ 2, 4, 6, 8, 10 . . .
$(O):$ 1, 3, 5, 7, 9 . . .

As before, we have only written the first few elements in each set, though each is endless. From these few, however, we see that we can tell which set or sets each number belongs to.

If we write $(E)+(O)$ for the set formed of all the elements of (E) and all the elements of (O), we find that $(E)+(O)=(I)$.

(E) and (O) are said to be two complementary sets in (I) in just the same way as 7 and 3 were complementary in 10 because $10=7+3$. Similarly, because we can find the difference between 10 and 7 and between 10 and 3, writing this as $10-7=3$ and $10-3=7$, we can operate upon (I), (E) and (O) in that way.

$$(I)-(E)=(O) \text{ and } (I)-(O)=(E)$$

6. Now, (E) could itself be written as the sum of two infinite sets, the first set being the doubles of the odd numbers and the second being the remainder of the even numbers:

2, 6, 10, 14, 18, 22 . . .
4, 8, 12, 16, 20, 24 . . .

So, instead of two sets forming (I) we can have three:

1, 3, 5, 7, 9, 11, 13, 15 . . .
2, 6, 10, 14, 18, 22, 26, 30 . . .
4, 8, 12, 16, 20, 24, 28, 32 . . .

None of these sets has any element belonging to another, and for this reason they are said to be *disjoint*.

Can we go one step further, writing the third set as two separate and disjoint infinite sets? The answer is *yes*, because we can first write down the doubles of the doubles of the odd numbers and then write down those even numbers that are left:

4, 12, 20, 28, 36, 44, 52, 60 . . .
8, 16, 24, 32, 40, 48, 56, 64 . . .

So (*I*) is now formed of four sets:

1, 3, 5, 7, 9, 11, 13, 15 . . .	the odd numbers
2, 6, 10, 14, 18, 22, 26, 30 . . .	their doubles
4, 12, 20, 28, 36, 44, 52, 60 . . .	their double-doubles
8, 16, 24, 32, 40, 48, 56, 64 . . .	the remainder.

But this last set can be formed of two disjoint sets, the first formed of the double of the double of the doubles of the odd numbers and the other of those left:

8, 24, 40, 56, 72, 88, 104, 120 . . .
16, 32, 48, 64, 80, 96, 112, 128 . . .

So that (*I*) is now formed of five of its parts, each being infinite and each being obtained by writing down the elements that are just double the elements of the previous set.

7. In fact, we can say, if we number the lines on which we write the sets: (*O*) for odd, (E_1) for the first even line whose elements are the doubles of the odd numbers, $(E_2), (E_3) (E_4)$ for the second, third and fourth doubling, that,

$$(I)=(O)+(E_1)+(E_2)+(E_3)+(E_4)+(R)$$

where (R) is the set formed by the remaining elements.

Can we go further, and write:

$$(I)=(O)+E_1+E_2+E_3+E_4+E_5+ \ . \ . \ .$$

where all the sets are formed by doubling the elements in the preceding set?

It is clear that we can, and if we form the beginning of a Table we see that it is endless both to the right and downwards:

(*O*)	1	3	5	7	9	11	13	15 . . .
(E_1)	2	6	10	14	18	22	26	30 . . .

(E_2)	4	12	20	28	36	44	52	60...
(E_3)	8	24	40	56	72	88	104	120...
(E_4)	16	48	80	112	144	176	208	240...
(E_5)	32	96	160	224	288	352	416	480...
(E_6)	64	192	320	448	576	704	832	960...
(E_7)	128	384	640	896	1152	1408	1664	1920...

.   .   .   .

.   .   .   .

.   .   .   .

Given any whole number we can find it in our table for, either it is odd and is somewhere on the first line, or it is even. If it is even, we can find the factor of it that is the highest power of 2 and, if it is, for example 2^7, we know that it is the double of the double of the double . . . (seven times) and must therefore be in the (E_7) line.

Think of any number and find out which line it is in.

Look at the table and see how the numbers on each line are spaced with respect to the sequence (I). This will help you to see that it is possible to go on adding new lines which contain numbers not previously met.

The richness of (I)

8. We can express what we have found in No. 7 in the following words:

The set (I) of integers can be split into an infinite number of parts each being infinite and equivalent to (I).

This is possible because (I) is itself infinite.

Now, if we look at the numbers in (E_{16}), for example, we can see that they begin with 2^{16} and the rest can be obtained by adding 2^{17} progressively as we move along to the right. Thus:

(E_{16}) 2^{16}, $2^{16}+2^{17}$, $2^{16}+2\times2^{17}$, $2^{16}+3\times2^{17}$, $2^{16}+4\times2^{17}$ . . .

Notice also how the numbers become larger and larger as we move downwards so that the whole numbers become more and more sparsely distributed. What is remarkable is that a very sparsely distributed set of whole numbers is equivalent to the original (I).

This can be clearly seen in the following examples:

(i) The set,

$$1, 2, 2^2, 2^3, 2^4, 2^5 \dots$$

is the set of the powers equivalent to (I), for we can write:

1	2	3	4	5	6	7 ...
$\updownarrow$	$\updownarrow$	$\updownarrow$	$\updownarrow$	$\updownarrow$	$\updownarrow$	$\updownarrow$
1	2	2^2	2^3	2^4	2^5	2^6

To every integer there corresponds a power of 2, and vice versa.

(ii) The set,

$$1, 10, 10^2, 10^3, 10^4, 10^5 \dots$$

is equivalent, similarly, to (I).

(iii) The set,

$$1, 100, 100^2, 100^3, 100^4, 100^5 \dots$$

and the set,

$$1, 10, 10^{10}, 10^{10^{10}}, 10^{10^{10^{10}}}, 10^{10^{10^{10^{10}}}}, \dots$$

(which is formed of numbers very thinly distributed among the integers), are still equivalent to (I) since a one-one correspondence can be established between the sets.

How to think about infinite sets

9. Now that we have begun to think about infinite sets, we can see that there are two ways of thinking about numbers. In one, we look for the properties of some numbers and, in the other, we consider sets and examine how they are defined by the properties of their elements, and what sets have a given property.

One of the questions we asked in Part III was: "Is the set of pairs of consecutive primes infinite?" We said that nobody yet knows the answer to that question.

Here is a question of this type which we *can* answer: "Is there a power of 2 in every group of 1000 consecutive numbers?" We can find the answer by doubling, and when we reach 1024 and double it (2048) we see that the gap exceeds 1000. If we proceed further we see the gap growing bigger: $2^{10}=1024$, $2^{11}=2048$, $2^{12}=4096$, $2^{13}=8192$ and $2^{14}=16384$

The powers of 2 are seen to become few and far between as we go on.

Here is another question which can be answered, but in which the answer is of a different type from the last and the question itself differs from all the questions we have put so far, presenting a quite new challenge: "Is the set of prime numbers infinite?"

We said that this could be answered affirmatively, but we cannot give that answer by making a trial as we did before because we cannot have an infinite set actually present to work upon. We use a method which is often used when we consider infinity; it is called *reductio ad absurdum* because we first assume that the contrary is true and then prove that this assumption cannot be right.

The set of prime numbers is infinite

10.   So, let us assume that there is a largest prime number, which we shall write down as P because we have only said that it exists and we cannot write it down in figures. P, being a number, can be multiplied by 2, then by 3, then by 4, then by 5, and so on, until we multiply it by the number immediately preceding it, namely $(P-1)$. We thus obtain a number which can be written:

$$1 \times 2 \times 3 \times 4 \times 5 \times 6 \times \ldots (P-2) \times (P-1) \times P$$

If to this number we add 1 we get an odd number which we shall call N.

$$N = 1 \times 2 \times 3 \times 4 \ldots \times (P-2) \times (P-1) \times P + 1$$

and is clearly bigger than P. The product on the right is divisible by P and by all the primes that are smaller than P and is therefore not a prime number. If N (which is this product plus one) were a composite number, and if P were the largest prime, the prime factors of N would certainly be found among the primes in the product on the right.

81

So $N - 1 \times 2 \times 3 \times 4 \times \ldots P$ would be divisible by these factors. But we already know that the difference between N and this product is 1 and as 1 has no other factor than itself, we must conclude that either N is prime and this means P is not the largest prime or it is composite and its prime factors cannot be smaller or equal to P. So again P cannot be the largest prime number.

In this way we have shown that if we suppose there is a largest prime, this supposition cannot be made to agree with what we can do with numbers. We are obliged, therefore, to abandon that supposition and to confess that we were wrong in supposing there was a largest prime. After every prime in the series of integers there are others. Indeed, there are as many as we want: there is an infinity of them.

11. To find them is, of course, a different matter. As soon as numbers contain too many figures we cannot grasp them with our minds. However, all the primes that are in the first few millions have been found so that we can know whether a number of 7 or 8 figures is a prime or we can find its prime factors.

Summary

12. In this section we have met the set of integers and have studied a few of its properties. We have also posed one or two questions about whether a given set (whose elements have a particular property) has a finite cardinal number *or* is infinite. We have seen in one case that we could decide this question and that in another case we could not. This shows that in mathematics not all questions are at the same level of difficulty even if they appear to be so.

We have learned to think in a variety of ways and more will be added as we go along. The intention is that you should expand your mental powers and come to know what your mind is capable of grasping.

You will have realised that your power has been greatly increased already by what you have learned. When you learned how to calculate you found you could save time and achieve in minutes or even seconds what might otherwise have taken hours or even days. In mathematics we are always trying to widen our field, to increase our capacity to see far and see clearly, and to do quickly what seems to require a long time.

When we think of infinity we show that we can gather up into one thought what would need eternity to be grasped. If we can do this, as we have tried to in this section by means of a few simple questions, we have moved to a higher level of thought and we have expanded our powers.

In your future studies you will encounter much more to confirm that you can grow in power, using your mind better and reaching out to ever wider fields.

www.ingramcontent.com/pod-product-compliance
Lightning Source LLC
Chambersburg PA
CBHW050614210326
41521CB00008B/1245